Looking Beneath the Surface of Agricultural Safety and Health

Dennis J. Murphy
Penn State University

Published by the American Society of Agricultural Engineers

For information, contact the American Society of Agricultural Engineers,
2950 Niles Road, St. Joseph, Michigan 49085-9659
Phone 269.429.0300 Fax 269.429.3852

Denise Sicking, Copy Editor
Peg McCann, Editor and Production

International Standard Book Number (ISBN) 1-892769-28-X
ASAE Publication 801M0303
LCCN Number 2002116268
Manufactured in the United States of America

Contents

Executive Summary

Production agriculture safety and health has been heavily influenced by the positive contributions of engineers, educators, writers, farm parents and others, whose drive and commitment to reduce the hazards and risks of farming often involved an emotional tie to agriculture as a way of life, and to a deeply felt personal conviction that farm work injury could and should be prevented. It might be said that agricultural safety and health has been "compassion driven" rather than "evidence" or "theory driven." Modern production agriculture (farming) is characterized by an assortment of operations that range widely in the diversity of products produced, how much is produced, and organizational structure. These same characteristics contribute to the difficulty in finding workable solutions to agricultural work injury because the diversity and the small-business scale of farming precludes one-size, single-approach, economically efficient solutions.

Agricultural safety and health evolved from within the philosophical and practical structure of the industrial safety and health movement in the early 1900s. Even so, agriculture was recognized to be a fundamentally different occupational enterprise because of: (a) a lack of uniformity and control of workplaces and work activities, (b) an overlap between home and work sites, (c) labor supplied by workers who were younger and older than in other occupations, and (d) relatively little governmental regulation of farm work hazards and risks. The early history of agricultural safety and health education emphasized the development and implementation of voluntary educational injury prevention programs and campaigns. These programs and campaigns were based on available agricultural injury statistics and the idea that "safety education" was the key methodological approach by which farmers and their families could find protection from the hazards encountered in farming. The literature suggests that little has changed in the philosophical approach to farm safety by traditional, or agricultural-based, farm safety and health professionals since the inception of organized farm safety programming.

The history of public health sector participation is one of documenting farm-related traumatic injury and occupational disease and illness, and bringing public pressure to bear on the problems and issues of farm work safety and health. The earliest article found that addressed an occupational agriculture hazard was published in 1916, and articles and activities increased steadily through the next few decades. The mid and late 1980s show a marked increase in concern about agricultural safety and health issues by the public health sector. This concern, along with a new national report with the provocative title of *Agriculture At Risk: A Report to the Nation,* attracted the attention of the U.S. Congress, and a new National Initiative in Agricultural Safety and Health was created. One part of this federal initiative directed NIOSH to convene a Surgeon General's Conference on Agricultural Safety and Health. The purpose of the conference was to raise consciousness, build coalitions, disseminate information, and encourage action. The National Initiative represented the most important public-policy shift concerning occupational agriculture safety and health since the OSHA reform legislation in 1976. The National Initiative has been implemented by NIOSH through a series of funded extramural and intramural programs. This initiative is directing an unprecedented level of funding and resource support toward agricultural safety and health, and has raised the importance (value) of farm safety and health as a program area at land-grant universities. The most highly and consistently funded National Initiative program component is the Ag Centers.

Historical and contemporary impediments to reducing occupational agriculture injury are identified as permeating and negatively impacting progress in agricultural safety and health in the United States. This discussion in Chapter 4 is organized around three main themes: the nature of farm work, program theory and evaluation, and surveillance data issues. These themes are related only in that they represent long-term and difficult problems to overcome. Some of the specific problems identified within these themes were first noted as many as 60 years ago. Some problems have emerged only in more recent years, while still others have never been discussed as formally as they are here. These challenges are interrelated and often feed into each other; perhaps this is why they are not easily addressed or resolved. Nevertheless, these are issues that need a systematic, long-term, holistic approach if we are to make major improvements in our capacity to significantly improve occupational agriculture safety and health.

Broadly based psychological theories, concepts and principles, such as attitude formation, attribution theory, concepts of self, operant conditioning, learning, risk perception and decision-making, have lead to a multitude of more specific theories and principles about why people think and behave as they do regarding personal and occupational safety and health, and methods for modifying those thoughts and behaviors. Cognitively oriented theories of human behavior dominate today's psychology and is the type being applied in agricultural settings. At the same time, industrial operations throughout the world are achieving significant injury reductions through the application of a form of traditional behaviorism. From the perspective of the *farm safety—risk paradox* (incongruence and disconnect between safety knowledge, attitudes and practices), a question that may be asked is, "Under what circumstances might cognitive and behavioral theories and models of human behavior be applicable to farm safety and health education?" Possible answers to this question are explored.

The literature shows that learning theory quotes many of the same founding fathers of behavioral theory. A consideration of learning theory from the perspective of agricultural work safety and health brings into focus four issues that can help to clarify important facets of learning theory and farm safety and health education. These are: (a) terms and definitions, (b) learning from experience, (c) adult learning theory, and (d) applicability of traditional learning theory to farm safety and health education. Traditional learning theory has little relevance to modern farm safety and health education practice. A more contemporary approach to learning suggests that: (a) no one viewpoint or explanation of behavior can fully account for learning, (b) learning occurs in a variety of settings with learning in one setting often affecting the learning that occurs in another setting, (c) learning is best experienced in settings that acknowledge the uniqueness of individuals, and (d) using a variety of instructional approaches enhances learning. Narrative psychology provides educators with a fundamentally different approach to farm safety and health education that overcomes some of the limitations of past approaches. It represents a cognitive constructivist approach to understanding and learning, and is just one new way of approaching education and learning.

All of the topics and subjects addressed in this review have a bearing on future directions and accomplishment for those working professionally for improved agricultural safety and health. The nature of the farming enterprise, the reluctance society has toward coerced interference with safety related behavior on the farm, and the instability of resources and direction that is inherent with voluntary and public appropriation of funds, preclude quick fixes or easy resolutions. Several suggestions and recommendations offered as a possible approach for addressing the many interrelated problems, challenges and issues identified in this review are:

1. There should be a single national agenda for agricultural safety and health.
2. A consensus process is needed for guidelines on how dollars for the National Initiative should be divided between surveillance, research, intervention, professional development, and information dissemination efforts.
3. Alternative structures for the Ag Centers concept should be explored.
4. A consistent, inclusive and ongoing agricultural work-injury surveillance program that provides reliable and valid estimates of injury rates by exposure hours at the national and state level is needed, along with timely dissemination of such data.
5. The agricultural safety and health profession should have available a more fully defined, developed, and supported set of competencies for its areas of practice.
6. Irrespective of specific professional competencies, professional development for practitioners is needed around the issues of surveillance and research, injury prevention methods, intervention evaluation effectiveness methods, behavioral and learning theory and principles, developmentally appropriate tasks for children who work on farms, and ergonomic solutions to workplace hazards and environments.
7. A consensus document to establish a nomenclature is needed for agricultural injury data.
8. A major effort is needed to gain national grass-roots support from farm parents, farm organizations, and rural communities for minimal, national standards of acceptable conduct for deliberate exposure of children under the age of 16 to farm work and farm work environment hazards.
9. The production agriculture safety and health movement could benefit from a process and forum that would allow experienced injury prevention professionals an opportunity for in-depth exploration of cultural, social, and political barriers and impediments to improved agricultural safety and health.

Acknowledgements

This review was conducted between September 2000 and July 2001 while I was with the Division of Safety Research, National Institute for Occupational Safety and Health (NIOSH), Morgantown, West Virginia, and on a sabbatical leave from Penn State University. Steve Olenchock, David Hard, and Nancy Stout helped to make this possible, and I am truly appreciative. Everyone in the Division of Safety Research was supportive of my work, and made me feel very welcomed. Several people reviewed various drafts of my review as it progressed, but no reviewers did more for me than David Hard, John Myers, and Steve Olenchock. The Morgantown librarians were also particularly helpful. They always quickly found even the most obscure and rather difficult-to-find references. Everyone should have a work environment as pleasant and supportive as I experienced.

Several professionals outside of NIOSH also helped with technical reviews. They were: Dick Beyer, Louisiana Farm Bureau; David Baker, University of Missouri; Brad Rein, U.S. Department of Agriculture; Tim Kelsey, Penn State University; Kelley Donham, University of Iowa; Barb Lee, National Farm Medicine Center; and Roy Buchan, Colorado State University. All reviewed bits and pieces as the work progressed, or supplied me with important information. This review is more accurate because of their help.

The biggest challenge, though, clearly was to those who agreed to wade through the last full draft and offer constructive criticisms that would help make this review a positive contribution to the literature of agricultural safety and health. These key contributors were: Steve Olenchock, PhD, Senior Scientist for Agricultural Research; David Hard, PhD, Ag Safety and Health Specialist; and John Myers, MS, Statistician (Health). All work at the Division of Safety Research, NIOSH, Morgantown, West Virginia. Non-NIOSH scientific reviewers included: Larry Chapman, PhD, Senior Scientist, University of Wisconsin-Madison; Henry Cole, EdD, Southeast Center for Agricultural Health and Injury Prevention, University of Kentucky; Richard Fenske, PhD, Director, Pacific Northwest Agricultural Safety and Health Center, University of Washington; John May, MD, Director, New York Center for Agricultural Medicine and Health, Cooperstown, New York; and John A. Shutske, PhD, Extension Safety Specialist, Biosystems and Engineering, University of Minnesota. This review is more objective and more scholarly because of their suggestions and comments.

— Dennis J. Murphy

1 Reasons and Motivations for Conducting This Review

The history of agricultural safety and health is marked by several characteristics that have muddied progress toward "a significant and measurable impact on...health effects among rural Americans." This was the stated goal when Congress established a National Initiative in Agricultural Safety and Health under Public Law 101-517 in 1990 (Public Law 101-517, Senate Report, p. 94). The National Institute for Occupational Safety and Health (NIOSH), a center within the Centers for Disease Control and Prevention, Public Health Service, U.S. Department of Health and Human Services, was given the charge to develop a sustained program that would improve the health and safety of agricultural workers and their families over time. While 10 years is not a long time, given the nature of the challenge, it is an adequate amount of time to reflect on progress made, examine barriers that impede and cloud present-day efforts, and identify challenges and opportunities for future progress.

Evidence suggests that injury rate reductions in agriculture, while still in a downward trend, may have improved modestly at best since 1990. Depending on the data source (see Injury Trends section below), the improvement in the death rate during the 1980s was considerably more than that of the 1990s, or it has stayed the same. What is clearer is that significantly fewer resources existed for agricultural safety and health in the 1980s than in the 1990s. On the surface, this seems incongruent: less reduction in injury rates with increased resources going toward the effort. However, a surface view can hide underlying cultural, social, economic and political matters that constrain, hinder and cloud understanding. Looking beneath the surface can shine a light on underlying influences obscured by conflicting values and beliefs.

The history of production agriculture safety and health has been heavily influenced by the positive contributions of engineers, educators, writers, farm parents and others. Their drive and commitment to reduce the hazards and risks of farming often involved an emotional tie to agriculture as a way of life, and to a deeply felt personal conviction that farm work injury could and should be prevented. It might be said that agricultural safety and health has been "compassion driven" rather than "evidence" or "theory driven." For instance, in a 1943 issue of Farm Safety Review (NSC, 1943), one article describes a schoolteacher's pioneering efforts in farm safety that lead to the establishment of the Tri-County Farm and Home Safety Council (Oneida, Herkimer and Madison counties in rural New York). This schoolteacher quit her teaching job and began devoting all her time to farm safety as a result of a fatal accident involving one of her graduating students. This is not unlike the story of the Farm Safety For Just Kids (FS4JK) organization, started in 1987 by Marilyn Adams after the fatal accident of her 11-year-old son, Keith, in 1986 (Adams, 1997). But can compassion and an emotional tie alone—no matter how deeply felt and widely shared—provide a sufficient and capable structure from which to change the status quo of agricultural safety and health?

Modern production agriculture (farming) is characterized by an assortment of operations that range widely in how much is produced and in organizational structure. The Economic Research Service (ERS) of the U.S. Department of Agriculture (USDA) has recently developed a typology of farming that better represents a more statistically reliable picture of American farming than have previous models (Offutt, 2000). The new typology is based on the occupation of farm operators and the sales class of farms combined. This typology has produced eight groupings of farming, five of which are considered "small family farms" (sales < $250K):

1. *Limited resource.* Any small farm with gross sales less than $100,000, total farm assets less than $150,000, and total operator household income less than $20,000. Limited-resource farmers may report farming, a nonfarm occupation, or retirement as their major occupation.
2. *Retirement.* Small farms whose operators' report they are retired (excludes limited resource farms operating by retired farmers).
3. *Residential/lifestyle.* Small farms whose operators report a major occupation other than farming (excludes limited resource farms with operators reporting a nonfarm major occupation).
4. *Farming occupation/lower sales.* Small farms with sales less than $100,000, whose operators report farming as their major occupation (excludes limited-resource farms whose operators' report farming as their major occupation).
5. *Farming occupation/higher sales.* Small farms with sales between $100,000 and $250,000, whose operators report farming as their major occupation.
6. *Large family farms.* Farms with sales between $250,000 and $499,000.

7. *Very large family farms.* Farms with sales of $500,000 or more.
8. *Nonfamily farms.* Farms organized as nonfamily corporations or cooperatives, as well as farms operated by hired managers.

In 1997, 60% of all farms were 180 acres or less; just over 80% of all farms sold less than $100,000 of agricultural products; individuals, families or partnerships operated 95% of all farms; and 50% of all farm operators had a major occupation other than farming (USDA, 1997).

American farming is also characterized by a wide diversity of products. For example, in addition to commonly recognized beef, dairy, hog, sheep and poultry farms, there are mule, llama, buffalo, mink, fish and bee farms. Common crops are corn, soybean, wheat, and hay crops, but less common ones are sod and bedding/garden plants. Added to these types of farm operations are orchard, nut, fruit, and vegetable farms. With a few exceptions (most notably California and Florida), farming in the United States is characterized by small family farm operations growing and producing a wide variety of agricultural products. These same characteristics contribute to the difficulty in finding workable solutions to agricultural work injury because the diversity and the small-business scale of farming preclude one-size, single-approach, cost-efficient solutions.

What This Review Includes

Careful scrutiny of past and current approaches and issues is the key to advancing the state-of-the-art in any field of injury prevention. After commenting on a related conference activity that took place during this review, the remainder of this chapter will examine agricultural work injury data to find where serious work-related injury decreased during the past 20 years. Depending on the data source, it will show either a leveling off or a staying the course of injury rate reduction in the past eight to 10 years. The second and third chapters examine histories and contemporary perspectives of the two major forces in production agriculture safety and health today: traditional agriculturalists and public health. This examination will reveal more similarities than differences, and examines some strategic decisions of the current National Initiative in Agricultural Safety and Health. A fourth chapter then inspects fundamental challenges for production agriculture safety and health. Some of these challenges have existed from the beginning of organized farm safety and some are more recent, however all are present today and impede progress toward safer production agriculture.

A fifth chapter then tackles the complex area of behavioral models and farm safety and health education. Contemporary approaches to farm safety and health education suggest moving in the direction of predictive behavioral models as a key guide in the development of educational interventions. As this chapter will show, there is considerable unsettledness in behavioral psychology, and some of this unsettledness has important ramifications for agricultural safety and health. A sixth chapter then examines learning theory and farm safety and health education. A feature of this chapter is the uniting of behavioral and learning theory with farm safety and health education. The last chapter brings together the many challenges that face professionals working toward safer production agriculture, offering some suggestions for the future.

State-of-the-Art Conference

In addition to this book, other information is evolving. A nationwide strategic planning conference, "Using Past and Present to Map Future Actions," was held March 2-3, 2001 in Baltimore, Maryland. Dr. Robert (Chip) Petrea, Extension Specialist, Agricultural Safety and Health, University of Illinois at Urbana-Champaign, conceived of the idea for this conference. Organizational help came from a national planning committee and financial support from a number of agricultural and public health groups and agencies. The conference focused on where the field of agricultural safety and health is now in comparison to where it was in 1988 when the "Agricultural Occupational and Environmental Health: Policy Strategies for the Future" conference was held. Also under consideration is the conference's subsequent document, *Agriculture At Risk: A Report to the Nation.*

The 2001 conference was structured to present a "state-of-the-art" review of agricultural safety and health along several specific subject matter lines (e.g., children, migrant workers, injury data, tractors, environmental issues, etc.). Papers developed for this conference are being used as a base of information during a consensus-building process to produce an updated national resource document, much like the original *Agriculture At Risk* report. Even though I participated in this conference as a member of the ad hoc committee with a goal of producing a state-of-the-art national resource document, a few important differences will be evident between that document and this book.

As a single author I am unconstrained by a consensus process that often requires compromise in choice of subjects, words, interpretations, suggestions and recommendations. My only goal is to communicate to other professionals. A consensus document requires participant buy-in. I am also able to delve deeper into issues that do not fall into traditional farm safety and health subject matter lines. This book is intended to be broad and to be issue driven, and will incorporate a long-term historical development perspective. The document from the "Using Past and Present to Map Future Actions" conference will be organized along more traditional subject matter lines and have a shorter timeframe. Finally, this book has been developed for agricultural injury prevention professionals exclusively, while the conference document is intended as a resource for both professional and public audiences. For these three reasons, I hope that my book will be more provocative and more philosophical than a consensus document. At the same time, this book may be more likely to be biased or otherwise limited. It is hoped that both ultimately will prove to be valuable and complement each other.

Injury Trends

This section presents injury trends during the past 20 years as a way of setting the stage for the chapters that follow. The most reliable source of fatal injury data since 1992 comes from the U.S. Department of Labor, Bureau of Labor Statistics' (BLS) Census of Fatal Occupational Injury (CFOI) database (CDC, 2001). Case identification in the CFOI system relies on death certificates and at least one other source document such as a worker's compensation report or a police investigation report. Other fatal occupational injury identification systems, such as those by the National Safety Council (NSC) and NIOSH's National Traumatic Occupational Fatality (NTOF), have used death certificates only (CDC, 1987). The CFOI system, with its use of two source documents for case identification, better captures occupational agriculture fatalities to older workers and teens than do death certificate-only systems (Murphy & Yoder, 1998). This is important in agriculture because older workers in particular (e.g., age 65 and over) constitute a large percentage of fatal occupational agriculture cases (Janicak, 2000; Murphy & Yoder, 1998). The NSC has adopted the CFOI work death figure since 1992 (NSC, 2000). Because the BLS and the NSC use the same employment figures, their work death rate statistic has been the same since 1992.

Fatal injury data for the agriculture, forestry, and fishing industries (US OMB, 1987) from 1980 through the latest available year has been summarized in Table 1.1 for the three common sources of national-level fatality data, and is shown graphically for each year in Figure 1.1 (NSC, 1981-2000; NIOSH, 2001a; US BLS, 2000). The data from the three sources cannot be directly compared to one another across all the years because of different methods of collecting and analyzing the data. Nevertheless, the table and graph depict possible trends in death rates during a 20-year period. Details regarding case definitions, analyses procedures, database limitations, etc., are identified in the referenced materials.

Table 1.1. Agricultural deaths, death rates, and percent-of-death-rate change by data source.[a]

Source	NSC			NTOF			CFOI		
Years	Deaths	Death Rate	D.R. % Change	Deaths	Death Rate	D.R. % Change	Deaths	Death Rate	D.R. % Change
1980	2,000	61		821	23.7				
1989	1,300	42	-31.1 [b]	683	20.2	-14.8 [b]			
1990				603	-18.0				
1992	[c]						779	23.2	
1997				537	-14.9	-17.2 [d]			
1999							770	22.5	-3.0 [e]

[a] Work deaths per 100,000 workers.
[b] between 1980 and 1989.
[c] began using CFOI numbers.
[d] between 1990 and 1997.
[e] between 1992 and 1999.

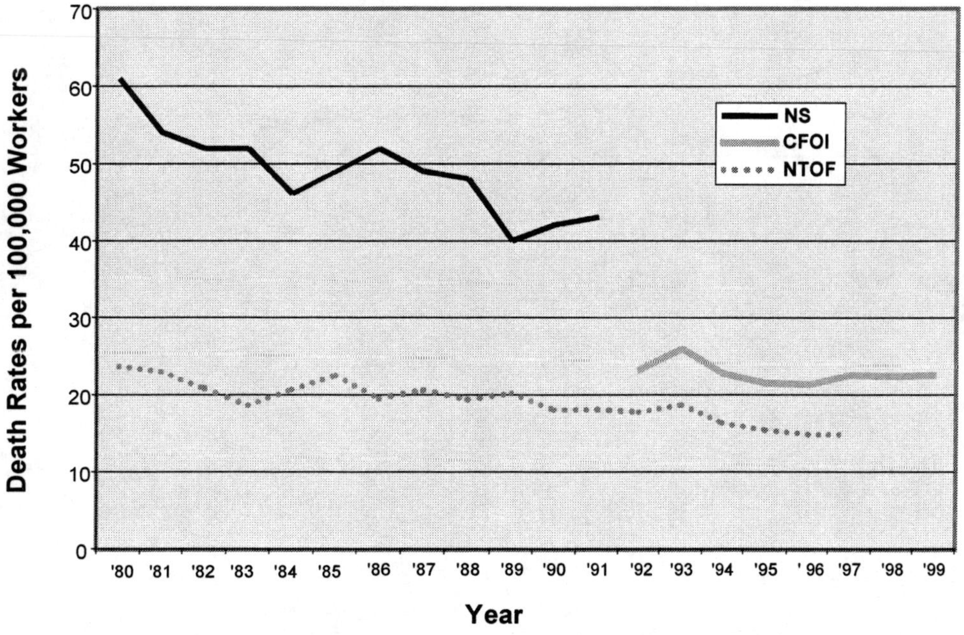

Figure 1.1. Agricultural death rates per 100,000 workers by year from NSC, CFOI and NTOF databases.

The data show that, while still improving, the pace at which the work death rate decreased has either slowed appreciably or has continued at about the same pace. It is the CFOI data that suggests the work death rate has changed little in the 1990s (3%), while the NTOF data suggests the work death has continued to decline at about the same rate in the 1990s (17.2%) as it did in the 1980s (14.8%). Regardless of the source, the data clearly suggests there has not been an accelerated reduction in the work death rate despite the unprecedented interest in, and financial support for, agricultural injury prevention. The chapters that follow this section examine facets of agricultural safety and health that may help explain this lack of progress.

To see if the agricultural injury trend differed from the all-industry trend, the NTOF database was analyzed for differences in the average annual fatality rate between 1980 and 1997. The all-industry fatality rates in 1980, 1989, 1990 and 1997 were, respectively, 7.46, 4.87, 4.57, and 4.1 fatalities per 100,000 workers. The percent decline among these rates is similar to the agricultural-only rates. That is, the all-industry average rate declined much more in the 1980s (34.7% from 1980 through 1989) than in the 1990s (10.3% from 1990 through 1997). Thus agriculture may simply be reflecting a trend across all industries.

The apparent slowdown in the reduction of work fatality rates may reflect the increasing difficulty in removing additional hazards from the work environment. That is, the initial improvements in fatality rates have come about by reducing or removing the worst hazards in the work environment. However, once these common hazards are controlled, additional hazard reduction becomes harder to accomplish or does not have as large an impact. Thus, it may be much more difficult to reduce fatality rates per 100,000 workers from 25 to 20, than it is to reduce these rates from 40 to 30.

Agriculture does not have a comprehensive source of annually published nonfatal injury data. The only national source of annually published nonfatal injury that allows trend analysis is the Survey of Occupational Injuries and Illnesses (SOII) data collected by the Bureau of Labor Statistics (BLS) as a part of their annual survey of private business establishments. The data comes from the Annual Log and Summary of Occupational Injuries and Illnesses (OSHA Form 200) as required by the Occupational Safety and Health Act of 1970 (Public Law 91-596). The major limitation of this survey as it pertains to agriculture is that farms with 10 or fewer employees do not have to complete the OSHA Form 200. Based on 1997 Census of Agriculture figures (Table 5), this excludes approximately 97% of farm operations from this database. Thus, the BLS SOII is only representative of larger agricultural operations.

Nonfatal injury data from 1980 through 1999 has been summarized in Table 1.2 and are shown graphically for each year in Figure 1.2 (US BLS, 2000). The rate of OSHA

Table 1.2. Incident rates of occupational injury and illness per 100 full-time agricultural workers from the U.S. Bureau of Labor Standards' Annual Survey of Occupational Injury and Illnesses.[a]

Year	Total Cases	% Change
1980	11.9	
1989	10.9	8.4 [b]
1990	11.6	
1999	7.3	37.1 [c]

[a] Excludes farms with fewer than 11 employees.
[b] From 1980.
[c] From 1990.

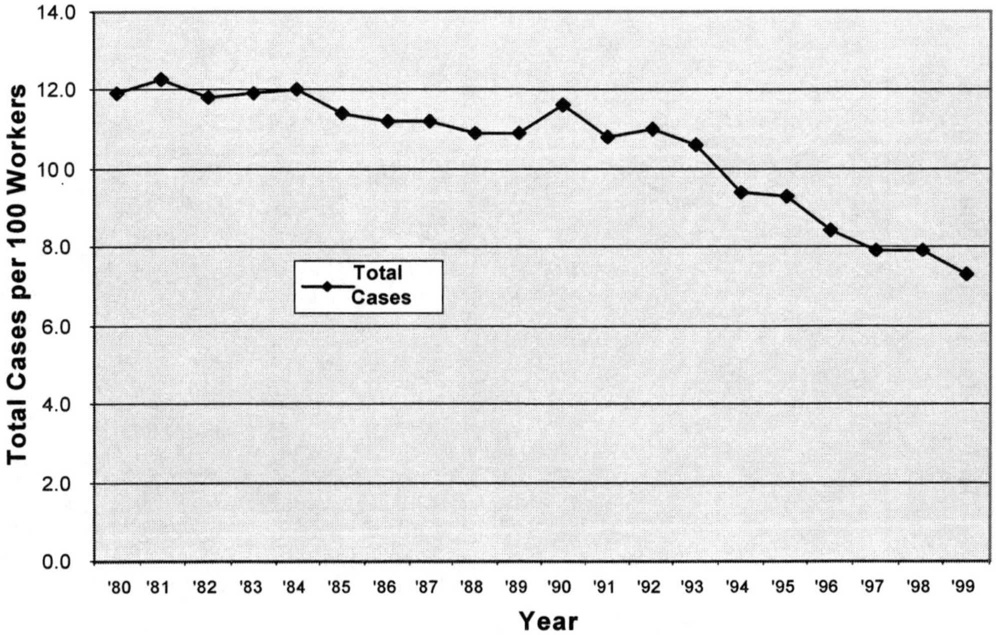

Figure 1.2. Annual incident rates of occupational injury and illness per 100 full-time workers from the U.S. Department of Labor, Bureau of Labor Statistics Annual Survey of Occupational Injury and Illness.

reportable occupational injury and illness cases in the BLS database declined from a high of 12.3 per 100 full-time workers in 1981 to 7.3 in 1999. The trend in the nonfatal injury rate differs from the fatality rate shown earlier by not only posting a large rate decline in the 1990s, but also one that is larger than the decline of the 1980s. One possible interpretation of this result is that because the nonfatal data comes from larger agricultural operations (a minimum of 11 or more hired workers), they are more likely to have active safety and health programs as compared to smaller family farm operations where family members supply most of the labor. While useful as a surveillance measure for large agricultural operations, the SOII cannot be considered a barometer of nonfatal injury for all of production agriculture because it excludes most agricultural operations. On the other hand, if all agricultural operations had active safety and health programs, the work injury rate for the entire industry may be reduced significantly.

Chapter Summary

Agricultural safety and health has historically been compassion driven rather than research or theory driven. Modern farm operations are characterized by their diversity in what is produced, with small family farm operations constituting the majority of all farm operations in the United States. Larger farms, while considerably fewer in number, produce the majority of agricultural products.

The most reliable and comprehensive source of occupational agriculture injury data, the Census of Fatal Occupational Injury, suggests that fatal agricultural work injury has not decreased appreciably during the past decade despite the significant influx of public health sector financial resources and expertise. Considerable caution is warranted in attempting to explain this phenomenon. Most importantly, it is not unreasonable to expect that any new initiative, such as the one undertaken by the public health sector for agricultural safety and health, may take a number of years to bear fruit for the bottom line—namely, a significant or even slightly accelerated reduction in the rate of worker deaths. Additionally, broad economic trends and a nonregulatory approach to agricultural safety and health must be considered in any hypotheses about why the agricultural injury death rate has leveled-off in recent years (CFOI data, Table 1.1). Nevertheless, whether it is rational or irrational, public policy makers may not wait much longer for solid evidence that the considerable investment of public dollars in agricultural safety and health is paying off in a significantly reduced rate of death.

On the other hand, and irrespective of public policy issues, professional ethics require efficient and effective protection for each and every life from the hazards of production agriculture. It is for all these reasons that agricultural safety and health professionals must continue to press for a common understanding of why production agriculture continues to experience a work death rate several times the national average for all industries, and what might be done differently in the future to change this reality.

2 Traditional Agriculture: Programs, People and Professional Development

This chapter describes how a particular approach to farm safety and health in the United States arose. This perspective comes from the "traditional agriculturalist" point of view. That is, it discusses farm safety and health as it became organized and grew into a national program as a result of efforts by persons and groups of people with first-hand knowledge and/or experience in farming. Although a previous history of farm safety and health has been developed (Murphy, 1992), this account focuses more on program development and the professionals who led the farm safety movement in its early years. This history is important as background and context for the chapters that follow.

Program Development

Historical documents clearly show that agricultural safety and health evolved from within the structure of the industrial safety and health movement. That movement began around 1906 and was formalized in 1913 with the founding of the National Industrial Safety Council, later shortened to the National Safety Council (NSC, 2000b). Farm safety educational materials can be found from as early as 1918 (USDA, 1918) and began appearing more frequently throughout the 1920s. But the movement for an organized approach to farm safety began formally in 1937 when a separate program for farm safety was held during the NSC's Safety Congress. During this meeting, a resolution was passed calling for the NSC to develop a "national farm safety program" (Zearing, 1947). Additional programs devoted to farm safety were held in succeeding years, and in 1941, the council established a new Home and Farm Safety Division. In 1942, the first "National Home and Farm Safety Conference" was held. In 1944, a separate Farm Division was formed and the first National Farm Safety Week was held.

Those pushing for a national farm safety program included: staff from within the NSC, several individuals from the USDA, agricultural engineers from land-grant universities, representatives of the farm equipment industry, the American Red Cross, the American Society of Agricultural Engineers, the Farm Equipment Institute (now evolved into the Association of Equipment Manufacturers), farm member organizations, representatives of various state government boards and agencies, the farm press, and representatives of industrial state-level safety councils (NSC, 1937-1938, 1940-1943). The common expertise, which these interested people and organizations brought to this mutual effort, was knowledge of agriculture and rural life.

By the time the farm safety movement became organized, several guiding principles had become firmly established within the industrial safety and health field (Murphy, 1992). For example, accidents "did not just happen, they were caused." According to this viewpoint, there were just two main causes: mechanical causes, responsible for about 25% of all accidents; and personal causes, which were responsible for about 75% of all accidents. The "Three Es" had become firmly established as the model for addressing industrial work hazards. Engineering directed management to design, or engineer, the work environment to eliminate or guard against accidents. Education directed management to educate and train workers in the use of machinery and equipment to prevent or minimize accidents. Enforcement directed management to set rules and regulations that could be enforced, such as wearing personal protective equipment.

Of the three Es, education was clearly thought to be the most important. Safety education's conceptual foundations were first articulated in the 1920s, and reflected industry's understanding of human safety-related behaviors at that time. In those days, safety education's task was twofold: to encourage self-protective behaviors, and to arouse emotional interest in safety. The purpose of arousing emotional interests was to convince workers that injuries could be avoided if only they weren't so careless and reckless in their behavior and ignorant of safety rules and regulations.

This was the climate in the late 1930s and early 1940s, from which the nationally organized program of farm safety emerged. At the same time, agriculture was recognized to be a fundamentally different occupational enterprise (NSC, 1937). There was recognition of:

1. A lack of uniformity and control of workplaces and work activities,
2. An overlap between home sites and work sites,
3. Labor supplied by workers who were younger and older than in other occupations, and
4. Relatively little governmental regulation of farm work hazards and risks.

Embracing voluntary safety education as the "key" to agricultural safety and health represented not only a sound philosophical approach, but also a very practical approach in that historical context.

Educational Program Foundation

A foundation for the educational approach to the national farm safety program was articulated as early as the very first farm safety meeting in 1937. J.C. Mohler, Secretary, Kansas State Board of Education, described the Kansas approach to farm safety when he said, "What is needed is to get the facts about accidents into the hands and heads and hearts of every farm family, an educational campaign that will reach every home. The question is, how can the individual best be educated and made accident-conscious and what are the most effective ways and means of accomplishing these ends?" (Mohler, 1937, p. 45). He identified the state's farm safety committee members as the State Superintendent of Public Instruction, the State Leader of 4-H Clubs, the State Supervisor of Vocational Agriculture, the Managing Editor of the state's leading farm journal, and the Secretary of the State Board of Agriculture. He also recommended an "anti-farm accident plan" [sic] for other states. The plan had nine steps:

1. A full-time paid state director.
2. A farm accident survey.
3. A prevention program based on best available information.
4. Organize rural youths to carry program to farm families.
5. Cooperation of farm organizations.
6. Teaching of safety in public schools.
7. Community schools of instruction in machinery operation.
8. Literature giving concise farm accident facts.
9. Publicity—press, radio, movies, meetings, signs posted conspicuously.

C.M. Seagraves, Director, Department of Safety, Illinois Agricultural Association, at the same meeting, reiterated most of these themes in his presentation titled, "A Program for Farm Safety" (Seagraves, 1937).

The early history of agricultural safety and health education emphasized the development and implementation of voluntary educational injury prevention programs and campaigns. These programs and campaigns were based on available agricultural injury statistics and the idea that "safety education" was the key methodological approach by which farmers and their families could find protection from the hazards encountered in farming.

The periodical *Farm Safety Review*, published between 1943 and 1986 by NSC, provides excellent historical documentation of the means and methods of organized farm safety education, particularly from its inception in 1943 through the 1950s. By 1959, 45 states had established state farm safety committees; 4-H clubs and Future Farmers of America chapters were regularly engaged in safety contests and campaigns; and a wealth of safety posters, filmstrips, movies and slide sets were developed and distributed. Industry, particularly machinery and equipment manufacturers, were providing financial support. County fairs, county extension meetings, farm member organization meetings, the local, state and national press, and special farm safety days and conferences were all regular venues for farm safety education.

Program leadership in most states during the 1950s and 1960s came from extension farm safety specialists, farm bureau safety leaders, or a combination of the two when both organizations had an identified specialist for farm safety. In particular state farm bureaus were active in adding safety leaders during this time. The first national meeting of state farm bureau safety directors was held in 1955 in St. Louis, Missouri, with nine states represented (NSC, 1955). In 1962, when the National Institute for Farm Safety (NIFS) incorporated, 63% of its charter membership was affiliated with farm bureau or cooperative extension (NIFS is discussed in more detail in the next section). Farm bureau and cooperative extension, assisted by a few machinery manufacturers and rural insurance providers, continued to be the dominant leadership group for farm safety and health and injury prevention programs into the 1990s.

Farm safety programs received a boost in 1975 when Congress appropriated funds for the Extension Service, USDA, to support a state farm safety program in each state ($15,000 per state). The program was to be implemented by hiring an extension safety program leader (i.e. a farm safety specialist) in states without an extension program leader, and augmenting programs in states where an extension safety leader was already in place (Kirby, 1975). Funds were increased to $20,000 per state in FY 1976. Documents developed in support of these funds show how the educational approach to farm safety bridged the early days of farm safety to at least the 1990s.

The effort to receive these funds can be traced to 1971 and a meeting between a delegation from the NSC and U.S. Secretary of Agriculture Clifford M. Hardin (NSC, 1971a). Recent passage of the Occupational Safety and Health Act (OSHA) and agriculture's increase in its work death rate (58/100,000 workers in 1960, 67/100,000 in 1970 [NSC, 1971b]) were two strong motivators for this meeting. The delegation encouraged the secretary to increase USDA support for a national program for farm safety. Subsequent actions included the establishment of a USDA Task Force on Safety in Agriculture (representing agencies within USDA) with a full-time executive secretary and a National Advisory Committee on Safety in Agriculture (Davis, 1971).

Program Development Revisited

In 1973, the Extension Service convened a national taskforce of farm safety leaders to develop "the fundamentals for an identifiable national farm safety program" (Gilden, undated, p. 1) under the direction of Robert O. Gilden, who had been hired as the USDA taskforce executive secretary. This plan spelled out several components of a national farm safety program. A paragraph in the plan described what they thought a good agricultural safety program leader should do:

> ...(1) uses organizational skills to get people to work together on the solution of common problems or

promotion of common endeavors, (2) disseminates accident prevention and loss control information, (3) identifies and appraises the agricultural accident problem, (4) develops accident prevention and loss control procedures, and (5) measures effectiveness of procedures... (p. 1).

The report acknowledged that these five points were derived from another document, *Agricultural Safety Leader: Suggested Functions, Duties, and Background.* This document had been developed in cooperation with the NIFS, the Farm Conference of the NSC and the American Farm Bureau. The Agricultural Safety Leader (ASL) document further articulated duties and background regarding the five points (identified as functions in the ASL document, see Appendix A). Furthermore, the taskforce report stated, "Generalized duties and background of the safety leader were developed for the use of safety leaders in farm organizations as well as Extension Services" (p. 2). By 1978, 48 states and Puerto Rico had extension safety program leaders (Tofany, 1978). The 1978 Farm Bureau Safety Director's Conference attendance roster listed 50 farm bureau participants from 19 states and the national office.

It is significant that in 1971, agricultural leaders were still calling for a "national farm safety program." Despite the plethora of state committees, safety pamphlets, brochures and fact sheets, movies, slide sets, and state and national campaigns, the perspective of agricultural and safety leaders was that a coordinated and comprehensive national farm safety plan was lacking. Burke (1987), in his historical overview of farm safety, noted that the late 1950s and early 1960s were the high-water mark for state and local safety activities. The similarity between the 1973 ideas for a national farm safety program and the ideas expressed back in 1937 by Mohler, Seagraves and others is apparent. A paid, full-time program leader, a cadre of voluntary cooperators, reliance on farm injury data, and development and dissemination of educational materials were key strategies and elements for a national farm safety program. These ideas were reconfirmed in 1982 by Mary Nell Greenwood, Administrator for USDA Extension, in a letter to State Extension directors regarding appropriate use of funds for state extension farm safety programs (Greenwood, 1983).

The issue of program guidelines for state extension farm safety programs was not officially revisited until 1998 when Brad Rein, USDA National Program Leader (Safety), reported on the recommendations of a taskforce on how the extension farm safety funds might be most-effectively allocated to states in support of state extension farm safety programs (Rein, 1998a). The level of funding per state had never been increased from the 1976 level ($20K), and in fact, had been reduced to approximately $18,500 per state by the Gramm-Rudman-Hollings Anti-Deficit Act of 1985. Available USDA cooperative extension records do not allow identification of states by status of program area, but by 1990 the number of states with full-time or part-time farm safety specialists, using membership in NIFS as a surrogate measure, had been reduced to 28 states (from a high of 48 in 1978). However, every state was still receiving an equal amount of farm safety funding. Thus, a primary issue for the 1998 taskforce was financial as well as programmatic. The report stressed the need for state specialists to form an advisory group to help support their state's farm safety education program.

As a result of the taskforce deliberations, a committee of State Extension Farm Safety Leaders was appointed in early 1998 to develop new reporting guidelines for state program plans (State Farm Safety Plan of Work) to help show the national impacts of the programs. The structure of the final reporting guidelines suggested that traditional program elements, such as safety committees, collecting farm injury data, developing educational materials and involving other agricultural organizations, were still major strategies (Rein, 1998b). One new idea expressed was that State Plans of Work include a strategy for developing resources to support priority issues within the state and to report what resources (financial) were generated from other sources.

Based on the discussion above, it is apparent that little has changed in the philosophical approach to farm safety by traditional, or agricultural-based, farm safety and health professionals since the inception of organized farm safety programming. There have been the inevitable ups and downs in programs, resources and activities, tied primarily to broad U.S. economic trends. Some hazards became less prominent, such as blasting caps and dairy bulls, while new hazards surfaced, such as all-terrain vehicles. However, the basic approach of educating farm people with the hope that they would voluntarily take actions to protect themselves, has stayed intact. Farm safety education has always consisted mainly of recommendations to remove or correct physical hazards, to behave or act in certain ways when faced with a hazardous situation, and to wear personal protective equipment or devices.

The Beginning of the Profession

Farm safety and health as a professional field of practice closely parallels development of farm safety as an organized cause. The beginning of professional farm safety and health practice is difficult to identify precisely. NIFS suggests that it began at a state level in 1943 with the hiring of Randall C. Swanson as the first full-time cooperative extension farm safety specialist (NIFS, 1982). As explained below, others had titles suggestive of professional farm safety practice prior to 1943, but it is not clear what their responsibility included. In practice, cooperative extension specialists in various subject matters are employed to lead statewide educational programs. Joining Swanson as a pioneering extension farm safety specialist was: in 1945, Wilbur E. Stuckey, Ohio State University; in 1946, F.R. "Dick" Willsey, Purdue University and Fred Roth, Michigan State University; in 1947, Norval J. Wardle, Iowa State University; and in 1949, Glenn Prickett, University of Minnesota.

Cooperative Extension is the third of three missions Congress has given to land-grant universities over the years. The general mission of Cooperative Extension, as stated in the 1914 legislation creating it, was to "...aid in diffusing among the people of the United States useful and practical information...and to encourage the application of the same...." Today this mission is stated as "...the development of practical applications of research knowledge and giving of instruction and practical demonstrations of existing or improved practices and technologies" (Seevers, 1997). In other words, Cooperative Extension's mission is education. In popular parlance, its mission is to "help people help themselves." A cadre of state-level specialists and county-based agents carries out the cooperative extension mission. These professionals work together to develop and implement educational programs for farmers and other clientele groups.

Several other people carried the title of farm safety specialist, or a correlate, during this time, although records do not establish precisely when such an appointment was made in many cases. The oldest record of a professional farm safety practitioner appears to be that of C.M. Seagraves of the Illinois Agricultural Association. The 1937 record of the 26th National Safety Congress lists Seagraves as "Director, Department of Safety, Illinois Agricultural Association." Seagraves was one of the original agricultural leaders calling for an organized program of farm safety. Records from the 32nd National Safety Congress show that the 1943 Farm Safety Section was presided over by Harry M. Pontius, "Safety Director, Ohio Farm Bureau and Farm Bureau Insurance Companies...." Maynard Coe became the first full-time director of the NSC's newly created Farm Division in 1944, and is credited with developing farm safety as a professional field of practice (NIFS, 1982).

The March-April 1945 issue of *Farm Safety Review* contains an article featuring the work of Katherine Olmsted, Field Instructor, New York State Department of Health (NSC, 1945). She lists the objectives of that department's program for farm safety, describes two "training institutes" held for farm safety, and identifies numerous organizations and groups actively supporting farm safety in 1944. The March-April 1946 *Farm Safety Review* introduced Ken Query as a Pennsylvania farm safety specialist with the Pennsylvania Chain Store Council (NSC, 1946). Finally, the 1953 program of the Farm and Home Education Division, Department of Public Instruction, Commonwealth of Pennsylvania, identifies Samuel L. Horst as having been full-time "Chief of Farm and Home Safety Education" since 1947.

Professional Development

In April 1945, Maynard Coe, as Director of the Farm Division, NSC, invited state farm safety specialists to Chicago to a two-day "institute for farm safety specialists...to bring inspiration and knowledge..." (Young, 1996). The first institute attendees were Randall Swanson, Katherine Olmsted and Wilbur Stuckey. The three original specialists, plus F.R. Willsey and Fred Roth, attended the second "institute for farm safety specialists" in 1946 in Chicago. Similar institutes were held in 1947 and 1948. In 1949, the now-named "Farm Safety Specialists Institute" was held in Madison, Wisconsin with 12 active participants. In addition to extension specialists, participants included state farm bureau safety leaders, NSC staff and others, including for the first time, a representative from a machinery manufacturer. That the meeting was in Wisconsin reflected the desire of the professionals to form their own group, separate from the Farm Division of the NSC. In 1950, when the group met at Purdue University for the sixth annual Farm Safety Specialists Institute, 16 people professionally involved in farm safety were present (Burke, 1987).

The annual gathering of farm safety professionals continued to gain momentum throughout the 1950s. From the 1949 meeting on, active members would "host" the conference, thus the meeting established a history of moving around the country. The group meetings changed names a few more times until 1956 when it settled as the "National Institute for Farm Safety" (Young, 1996). At the July 1962 meeting in Florida, members voted to incorporate as a nonprofit, unaffiliated professional organization for farm safety leaders, and thus became the National Institute for Farm Safety, Inc. According to the incorporation papers filed with the state of Wisconsin, its purpose was "to advance the professional status of the career worker in farm safety and to promote the farm safety movement in the United States (Certificate of Incorporation, Department of State, Wisconsin, 1962). There were more than 60 professionals from 25 states, Puerto Rico and Canada at the 1962 meeting. By this time not all participants at the annual meetings worked full time in farm safety. Rather, some attendees had farm safety assigned to them by their employer as one of many duties, and others with varying levels of interest would attend when a meeting was held close to their home base. Examples of this latter group included local, state and national farm press, farm safety committees, and professional and government groups and agencies.

To ensure that anyone with a professional interest in farm safety, and in the organization, felt welcomed to join the new organization, a "charter" membership was offered to all those so desiring it (Young, 1996). The charter member offering was for one year until September 1, 1963. The result was that 46 individuals became Charter Members of NIFS. Twenty-two people were classified as "Members" while the other 24 were classified as "Associate Members." The difference between these two groups was that the Members category was for those professionally devoting a majority of their time to farm safety and who had done so for a minimum of five years, while the Associate Members category was for those devoting less time or with less than five years service to farm safety (NIFS, 1962). While other member categories were also created, e.g., Sustaining Members and Honorary Members, or have been since been added, e.g., Life Member, and individuals could be in more than one member category, the Member and Associate Member categories best reflect the professional expertise and number of members within NIFS.

The employer affiliations of these charter members provide a good picture of farm safety leaders at this point in time. Knowing these affiliations helps to explain the continuity of traditional safety education techniques as the major methodological approach to farm safety and health from its inception in the 1930s and 1940s through the 1990s. A membership roster with employer affiliation was included with the 1966 NIFS meeting proceedings. Between cross-referencing this list with the names of charter members, and examining other documents, employer affiliation was found for 45 of 46 charter members. These affiliations were: state Farm Bureaus – 16, state Cooperative Extension Services – 13, insurance companies – 5 (four were with one company), machinery manufacturers – 2, and assorted others representing a wide range of groups, organizations and employers interested in agriculture safety (for a complete list, see Appendix B).

As noted in a NIFS brochure marking the 10th anniversary of the organization, these pioneers "represented almost all of the farm safety expertise of the nation" and "their responsibilities by and large were in the field of education" (NIFS, 1973). The brochure also states that representatives of machinery manufacturers joined in large numbers in the 1960s, largely in response to an "Engineering Seminar," sponsored jointly by NIFS and the Farm and Industrial Equipment Institute (Association of Equipment Manufacturers) in 1966. Power take-off safety and rollover protection for tractors were the primary topics of the day-long seminar. While their number has increased steadily over the years as a percent of membership, the machinery and equipment industries have never constituted a large membership segment of NIFS.

Membership in NIFS reflects not only the growth of farm safety and health as a professional field of practice, but also its increasing diversity over the years. For example, as shown in Table 2.1, farm bureau and cooperative extension made up 63% of membership in NIFS in 1962, and there were no members with a public health affiliation. By 1990, farm bureau and cooperative extension constituted less than one-half the member affiliations (47%) and the public health category comprised 6% of the member affiliations. In 2000, farm bureau and cooperative extension comprised 35% of the member affiliations and public health had grown to 18%. Increased diversity is also revealed in growth of the "other" affiliations category as well, from 22% in 1962 to 28% in 2000. Finally, in 1962 there were two Canadian members with no other countries represented, while in 2000 there were five Canadian members and a member each from Finland and Sweden.

While the membership makeup of NIFS has changed in recent years, the important point here is that historically NIFS has been the organization where professionals, working either full time or nearly full time in the interest of farm safety and health, have come together for the purpose of furthering their professional development and "to promote the farm safety movement in the United States" (NIFS, 1962). As explained earlier, the "farm safety movement" primarily meant farm safety and health education. With only a few exceptions, early cooperative extension safety specialists and Farm Bureau leaders did not have academic degrees in education, occupational safety or health, public health, or closely allied fields. Nor was there an expectation of formal evaluation of extension or farm bureau safety education programs. Learning how to be a farm safety specialist (educator) or leader most often came about on the job, through reviewing published resource materials, by attending professional conferences and workshops, and through personal contacts with more experienced specialists and leaders.

A byproduct of this method of learning and indoctrination to job responsibility has been that traditional methods of program development and implementation were passed on, often without critical examination of program effectiveness. Most of these pioneers had first-hand experience with farming practices, thus the content of educational programs and campaigns was largely intuitive and practitioner driven. In keeping with the accepted view of occupational safety and health education as subject matter to be taught by a safety expert (teacher) to a worker (student), programs and education were oriented toward individual attitudes and behavior and relied heavily upon general safety rules and recommendations. As will be shown in the next chapter, this approach to farm safety and health was not questioned until the 1980s.

Table 2.1. Membership in NIFS by category in selected years from 1962 through 2000.

				Employer Affiliation by Percent of Total											
				Farm Bureau		Coop. Extension		Insurance		Machinery, Equip. Ind.		Other		Public Health	
Year	Member	Associate	Total[a]	No.	%	No.	%	No.	%	No.	%	No.	%	No.	%
1962	27	19	46	16	35	13	28	5	11	2	4	10	22	0	0
1970	29	58	87	35	40	20	23	7	8	7	8	18	21	0	0
1980	35	105	140	45	32	45	32	10	7	10	6	29	21	1	1
1990	50	89	139	28	20	37	27	15	11	13	9	38	27	8	6
2000	71	93	164	21	13	36	22	14	8	18	11	46	28	29	18

[a] Excludes categories of Honorary, Sustaining, and Life, and members listed as retired or members with missing affiliation data (6 in 1970, 19 in 1980, 23 in 1990, 20 in 2000).

Chapter Summary

This chapter has looked broadly at the development of agricultural safety and health as a professional field of practice by agriculturalists. The national approach to agricultural safety and health was heavily influenced by the industrial safety and health movement of the early 20th century. The industrial safety and health movement influenced agriculturalists to approach farm safety hazards, risk and injury reduction from an educational program perspective. That is, the problems of farm work safety and health could be resolved through voluntary educational programs. The first full-time professional practitioners were largely state cooperative extension agricultural engineers and educators, state farm bureau representatives, and farm insurance company representatives. The most common background of all was their tie to agriculture through practical experience. These agriculturalists formed the NIFS to further their professional development and to promote the farm safety movement. The basic structure of a national program for agricultural safety and health, as lead by agriculturalists, has not changed appreciably since its inception. This approach was not seriously challenged until the 1980s. Interestingly, as will be shown in later chapters, those challenging the voluntary educational approach taken by agriculturalists have largely come to adopt methods similar to the agriculturalists.

Appendix A. Agricultural Safety Leader (ASL) Document

AGRICULTURAL SAFETY LEADER

Suggested Functions, Duties, and Background

FUNCTION	DUTIES	BACKGROUND
I. Organizational skills	• Develop relationships with mutually concerned organizations. • Develop a cohesive working force within the geographic area of responsibility. • Develop techniques to gain multiplying effects of educational programs. • Develop policies, codes, safety standards, procedures, and educational programs designed to assist and educate rural people to prevent accidents.	• Ability to motivate others. • Ability to work with persons from many disciplines. • Knowledge of and ability to develop educational activities and programs.
II. Dissemination of accident prevention and loss-control information	• Utilize communication channels directed toward rural people. • Seek involvement in safety programs by persons and groups who serve agriculture.	• Ability to administer and carry out procedures and educational programs. • Thorough knowledge of communication channels and techniques used to reach the rural resident.
III. Identification and appraisal of the accident problem	• Develop statistical procedures and the conduct of studies that will identify the accident problem and determine trends. • Interpret accident information to determine needs and priorities. • Review the man-machine-environment system to identify likely accident producing situations. • Initiate studies of specific agricultural safety problems.	• Knowledge in physical and social sciences such as statistics, physics, chemistry, fundamentals of various engineering disciplines and human behavior. • Ability to apply the principles of measurement and analysis.
IV. Development of accident prevention and loss-control procedures	• Prescribe an accident and loss-control system designed to: 1. Eliminate causative factors associated with accidents. 2. Reduce the degree of hazard. 3. Reduce the severity of accidents. • Act as consultant to personnel engaged in the planning, design, development, and installation of various systems and procedures on the farm. • Systematically maintain current knowledge of human and technological developments, which may affect the safety profile.	• Specialized knowledge of accident causation and control. • Ability to administer and carry out procedures and programs.
V. Measurement of effectiveness of procedures	• Develop techniques for information retrieval to assist in evaluating programs and trends. • Develop techniques to evaluate countermeasures prescribed to determine effectiveness. • Develop techniques to evaluate human and cost benefits derived from a program.	• Knowledge of information retrieval systems. • The ability to apply various statistical and analytical techniques to evaluate both technical and human responses.

Appendix B. Charter Members and Employer Affiliations of National Institute for Farm Safety, Inc., 1962

Edward Adams, Iowa Farm Bureau

Orville Asper, Country Companies Insurance

Orville Barbee, Iowa Farm Bureau

Orrin Berge, Agriculture Engineer, University of Wisconsin

Herb Bogart, Industrial Commission, Ohio

Wendell Bowers, Agriculture Engineer, University of Illinois

James Burns, Louisiana Farm Bureau

George Cappe, Florida Farm Bureau

U. W. Caves, Louisiana Farm Bureau

Maynard Coe, National Safety Council

Ward Cross, Country Companies Insurance

Carroll Edcouch, Florida Farm Bureau

Emil Dietz, California Farm Bureau

Delbert Ekart, Kansas Farm Bureau

Ray Forsythe, Iowa Farm Bureau

Russell Heston, Grinnell Mutual Reinsurance Co.

Thomas Hippler, Louisiana Farm Bureau

Ordie Hogsett, Agricultural Engineer, University of Illinois

Robert Holmes, undetermined

Allen Humphrey, Iowa Farm Bureau

Donald Jensen, Agricultural Engineer, University of Wisconsin

E. Cullen Johnson, Virginia Farm Bureau

Eugene Lemmon, Farm Safety Council, Ontario, Canada

John Marks, Automotive Safety Foundation

James McAlister, Extension Specialist, Oregon State University

J.J. Messerschmidt, Automotive Safety Foundation

Ralph Patterson, USDA Cooperative Extension Service

A.M. Pettis, Extension Safety Leader, University of Florida

Glenn Pickett, Agricultural Engineer, University of Minnesota

Hugh Rhodes, Indiana Farm Bureau

John Rush, USDA Agricultural Research Service

Rollin Schneider, Agricultural Engineer, University of Nebraska

Grice Sexton, Kansa Farm Bureau

Joseph Smetana, Texas Farm Bureau

Willis Smith, Farm Safety Services, Bloomington, IL

Clarence Steven, Agricultural Engineer, University of Missouri

Wilbur Stuckey, Agricultural Engineer, Ohio State University

Randall Swanson, Agricultural Engineer, University of Wisconsin

Raymond Talbert, Iowa Farm Bureau

Edwin Tanquary, International Harvester, Hinsdale, Illinois

Norval Wardle, Agricultural Education, Iowa State University

James Williams, Country Companies Insurance Co., Illinois

F.R. Willsey, Agricultural Engineer, Purdue University

Daryll Wilson, Country Companies Insurance Co., Illinois

Hal Wright, Ontario Department of Agriculture, Ontario, Canada

Carlton Zink, Agricultural Engineer, Deere and Co., Moline, Illinois

3 Public Health Involvements and Impacts

This chapter reviews both history and current medical and public health involvement in agricultural safety and health. While current activity (e.g., from the 1990s and on) is well documented, little literature was found that provides a historical account or context for the early years of involvement by physicians or public health groups, or connects past history to current activity (hereafter, for simplicity, the term public health will include doctors and the medical community). Articles by Donham and Mutel (1982) and Schenker (1996) are exceptions. The history of public health sector participation is one of documenting farm-related traumatic injury and occupational disease and illness, and bringing public pressure to bear on the problems and issues of farm work safety and health. This chapter will focus primarily on the documented record of public health sector involvement in the farm safety and health movement. Suggestions regarding a few problematic questions and issues raised in the course of this chapter are addressed more fully in later chapters. First, for readers less familiar with public health's approach to disease and injury prevention, a few brief comments about epidemiology and injury epidemiology are offered.

Epidemiology

The field of public health uses the scientific method of epidemiology to study disease, disability and death in human populations (Timmreck, 1994). The word "epidemiology" derives from the Greek root words: *epi* for among, *demos* for the people, and *logos* for thought or study of. Put together, epidemiology means the study of what is happening among the people. The science of epidemiology arose out of the study of infectious disease involving the interactions of three main factors: host, agent and environment. Modern epidemiology includes the study of noninfectious disease and injury. The origin of epidemiolgy is often traced back to London physician John Snow and his work with the cholera epidemic in 1853, although it can be traced back to Hippocrates (~460-377 B.C.). Timmreck (1994) suggests that a modern view of epidemiology includes a "time" factor and uses broader terms than the original host, agent, and environment. He includes these ideas in an updated version of the epidemiological triangle (Figure 3.1). Descriptions of modern epidemiology often reference it as a study of the distribution and determinants of health-related states or events in specified populations, and the application of this study to control of health problems (Last, 1995). In-depth details on the origins and development of epidemiology go beyond the scope of this review and are perhaps unnecessary for most readers. The text by Timmreck provides a good overview of epidemiology for those who may need a refresher.

Injury Epidemiology

Conceptually, the public health community views injury as a problem that is analogous to disease (Godfrey, 1937; Division of Medical Sciences, National Research Council, 1966; Committee on Trauma Research, National Research Council, 1985; Christoffel and Gallagher, 1999). When epidemiology is applied to the problem of injury it is normally referred to as "injury epidemiology." While "injury as a disease" is not a particularly apt metaphor because of fundamental differences between disease and injury (e.g., disease is unwanted while injury often results from things wanted), it has provided the public health community with a familiar and comprehensive framework for attacking injury problems.

Injury epidemiology traces its history to the works of DeHaven (1942), Gibson (1961), and Haddon (1963, 1970, 1980). Four principles of injury epidemiology that are particularly relevant to public health's approach to agricultural safety and health are:

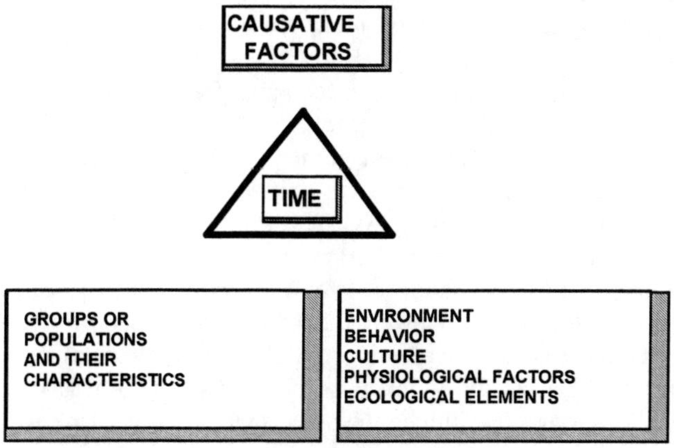

Figure 3.1. Advanced model of the triangle of epidemiology. Source: Timmreck TC [1994]. *An Introduction to Epidemiology,* p. 11.

1. An understanding of relevant person, place and time variables serve as the starting point in injury epidemiology.
2. Unintentional injuries are the result of excessive energy exchanges between people and energy sources. The problem of injury prevention is one of preventing and controlling these exchanges.
3. The selection of a preventive or control strategy is not dependant upon the rank order or importance of casual factors.
4. The prevention of injury does not depend upon an exhaustive understanding of injury causation.

The Committee on Trauma Research, National Research Council (1985) and Waller (1985) were among the first to develop texts on injury epidemiology, while Robertson (1998) and Christoffel and Gallagher (1999) provide more recent reference books. Murphy (1992) provides an overview of injury epidemiology and its connection to agricultural safety and health. Readers not familiar with the concepts of injury epidemiology are encouraged to review the last three citations.

The Early Years

Little is known about the earliest papers in the medical and public health literature so a database literature search was used to identify early medical and public health-based articles on agricultural safety and health. *Index Medicus* was searched for articles between 1917 (Vol. 1, 1916 was not available) and 1939, MEDLINE was searched for articles from its beginning in 1966 to 1980, and NIOSHTIC was searched for all years before 1980. An explanation of these databases can be found at www.nlm.nih.gov/pubs /factsheets/factsheets.html (*Index Medicus*, MEDLINE) and at www.cdc.gov/niosh/database.html (NIOSHTIC).

The earliest article found that addressed an occupational agriculture hazard was published in 1916. The article, by C.J. White, M.D., Professor of Dermatology at Harvard University, reviewed factors contributing to occupational skin diseases and noted that farmers were an occupational group associated with "excessive light" (quotations added). A.C. Borders, a physician with the Mayo Clinic, Rochester, Minnesota, similarly published two articles identifying farmers as an occupational group with high incidence of skin cancers (Borders, 1920, 1921).

Interestingly, the first published paper found in the public health literature that (a) primarily focused on occupational health or safety about farmers, and (b) specifically addressed traumatic injury, was written by E.L. Lloyd in 1933. He was listed as an Assistant Professor of Business Administration, Oklahoma Agricultural and Mechanical College (now Oklahoma State University), Stillwater, Oklahoma. Lloyd reviewed the causative agents of 35 agricultural accidents entering the University Hospital of the State University of Iowa (now University of Iowa) in 1931. Among other things, he noted that agricultural injuries were more severe than other types of industrial injuries treated at the same hospital.

A paper by Dr. M.O. Oppegaard, a surgeon from Crookston, Minnesota, was read at the Symposium on Emergency Surgery before the annual meeting of the Minnesota State Medical Association on May 24, 1933, and was published in March 1934 (Oppegaard, 1934). The article reviewed six cases that required emergency surgery as a result of farm accidents involving machinery and animals. Two interesting points made by Oppegaard before presenting the case histories were on progress in recent years with emergency surgery in rural areas. Oppegaard noted: (a) the advantages of being able to use a medical center for surgery rather than the farm kitchen, and (b) farm people were now willing to bring patients to the doctor rather than the doctor having to travel to the patient. In preparation for his paper, Oppegaard also stated that he enlisted the services of a national research service to obtain statistical information on his subject (farm injury). The national research service reported, "...a careful search of all available material failed to reveal any literature dealing specifically with farm accidents" (p. 128). The research service was not identified.

In 1939, Dr. John Powers summarized the types and consequences of farm injuries he had treated as a surgeon in the *Journal of the American Medical Association.* Powers practiced medicine at the Mary Imogene Bassett Hospital, in rural Cooperstown, New York. His paper was based on 310 farm accidents treated at the hospital between 1929 and 1938 (Powers, 1939). Importantly, Powers noted that, "Many farmers are poor, their equipment is inadequate and frequently out of repair, wages are low and hired help is difficult to find and often incompetent" (p. 1375).

In 1942, Dr. Kennedy Creevey, a physician practicing at the Mary McClellan Hospital in rural Cambridge, New York, developed a report on 370 cases of farm injury treated at his hospital between 1920 and 1940. Creevey, using the same methods and procedures of analyses as Powers did, noted many of the same basic causes that were identified in Powers' report but went on to provide a summary of his findings. In the summary, Creevey (1942) noted, among other things, that: (a) farming was naturally a hazardous occupation with hazards increased by poverty, which entailed poor training and poor equipment; (b) accident prevention as applied to industry is not easily adapted to farming, but improvement might be possible through *education* (emphasis added); and (c) suggestions for decreasing the accident rate consisted chiefly of the application of *care and common sense* (emphasis added).

In 1949, R.A. Calandruccio and Powers presented a paper that incorporated farm accident cases from the 1939 Powers report, some cases from the 1942 Creevey report, and all cases between 1939 and 1948 treated at the Mary Imogene Bassett Hospital in Cooperstown. This brought their total cases to 1,028. In a discussion following the report, it was noted that, "The farm tractor has been a factor in many of these injuries, due mainly to the carelessness of the individual farmer and also due to the failure of the manufacturers to equip these machines with proper safety devices" (p. 658), and "Interestingly enough, the hazards of industry, of mining and construction have been recognized for many years, and definite measures have been adopted for the promotion of safety. In agriculture no such progress has been made…" (p. 660).

It is significant that the first several public health sector articles on what we now call agricultural safety and health were by physicians—with the exception of Lloyd. As indicated by Powers and the other authors cited, the foundational understanding of the causes of farm accidents and what might be done about them was the same as those expressed by early agricultural safety professionals in their papers (e.g. see NSC, 1937-1943). Unlike traditional agricultural safety professionals though, it was not within the normal professional purview of physicians to devise or apply strategies and interventions to prevent farm work injury. Rather, these physicians, and the early epidemiologists, industrial hygienists, and other public health specialists that followed, concentrated on describing exposures and the circumstances and effects surrounding the event of interest. They searched for the who, when, where, what, how, etc., of farm work-related disease and injury. This is, of course, modern-day descriptive epidemiology.

Published papers and conference proceedings by public health-based professionals increased steadily each decade during the 1950s, 1960s, and 1970s. For example, the NIOSHTIC database found 34 English language citations on agricultural safety and health topics from 1950-1959, 116 citations from 1960-1969, and 341 citations from 1970-1979. Similarly, MEDLINE found 86 citations from 1966-1969 and 119 from 1976-1979. Notably, an examination of the public health literature on agricultural safety and health from the 1930s through the 1980s suggested that the public health sector did not move much beyond descriptive epidemiology (the field of pesticide safety is an exception to this statement). And in instances where public health professionals did actively engage in disease or injury prevention programs, they used the same educational methods (e.g. group meetings, films, publications) as did traditional agricultural safety educators. As is often said in the literature, this method has not been found to be a very effective strategy (e.g., Christoffel & Gallagher, 1999; Robertson, 1983). This may be why, from public health groups, one could still find in 1989 declarations such as, "America's most productive work force is being systematically liquidated by an epidemic of occupational disease and traumatic death and injury…" (Merchant et al., 1989, inside cover page) or the lack of rollover protection on farm tractors referred to in 1991 as an "occupational obscenity" (Myers et al., 1992, p. xviii).

The Beginning of Change

The mid-to-late 1980s show a marked increase in concern about agricultural safety and health issues by the public health sector. The groundwork for much of this activism was developed throughout the 1970s and into the mid-1980s by industrial hygienists, physicians with respiratory disease expertise, and researchers who were examining agricultural safety and health from an environmental health perspective. For example, in 1974, the Society of Occupational and Environmental Health sponsored a "Conference on Health and Safety in Agriculture" (SOEH, 1974), the University of Saskatchewan Respiratory Disease Clinic sponsored a 1977 international symposium on "Occupational Pulmonary Disease: Focus on Grain Dust and Health" (Dosman and Cotton, 1980), and the American Conference of Governmental Industrial Hygienist (ACGIH) sponsored an "Agricultural Respiratory Hazards" conference in 1982 (Kelley, 1982).

Organizational Developments

By the early 1980s, four distinct public health entities existed in the United States for the expressed purpose of conducting research and education on agricultural safety and health hazards and risks. These groups emerged out of ongoing work at their respective institutions because they saw both the need and opportunity to better marshal financial and human resources for their work in agricultural safety and health. Some of this work had been ongoing since the 1950s. Very similar to the many traditional agricultural groups, there was always considerable formal and informal collaboration among the professionals that staffed the four entities described below.

The first to be formed was the Institute of Agricultural Medicine (IAM) at the University of Iowa. The IAM was formed in 1955 as a part of the Department of Preventive Medicine and Environmental Health in the College of Medicine. The department head, a pediatrician named Franklin Top, and the dean of the college, Norman Nelson, M.D., collaborated with Dr. Mathew Kindee, Director of the Agricultural Division of the Kellogg Foundation, to develop the general concept of the institute as a body to study the specific health and injury problems of people working in agriculture. No such organization was in existence anywhere in North America at this time. The Kellogg Foundation, the philanthropic arm of Kellogg, Inc. (the cereal manufacturer), provided the initial funding. While Top, Nelson and Kindee are credited with the original concept and funding proposal, Clyde Berry, PhD, Industrial Hygiene, is credited with being the leader who converted the concept into a functional program (Donham, 2000).

The initial focus of the Iowa IAM was on finding and describing agricultural environmental and occupational hazards. Specialists with expertise in zoonoses, toxicology, agricultural anthropology, environmental health, and engineering were among the first employed in the late 1950s and 1960s, and other expertises, such as veterinary sciences and pulmonary medicine, were added later on. Major lines of research included protection from tractor rollover, power take-off guarding, lawn mower safety, pesticide exposures and health effects, agricultural industrial hygiene, agriculture respiratory hazards and mental health. The IAM became the Institute for Rural and Environmental Health (IREH) in 1998 and then in 1999 became the Department of Occupational and Environmental Health in the newly formed University of Iowa College of Public Health.

A sister organization of the IAM, the Institute of Rural Environmental Health (IREH) was formed in 1969 at Colorado State University, also by a Kellogg Foundation grant (Savage, 1974). Dr. John Collier, a veterinarian and professor in the Department of Microbiology, developed the proposal for the institute. Once funded, Dr. John Bagley, a retired Assistant Director for the Centers for Disease Control, developed and directed the institute. The institute's initial focuses were on relationships between high levels of nitrates in water and human and animal health, carbon monoxide poisonings in rural homes, pesticide exposures and health effects, and injuries involving pleasure horses. Eldon P. Savage, PhD, with expertise in pesticides, along with Bagley, established Colorado's first baccalaureate degree program in environmental health in 1971 and the Community Pesticides Center in 1972. Another major initiative of the IREH was a demonstration project to provide industrial hygiene and occupational safety services to small business (<100 employees) and agriculture. Dr. Roy Buchan, an Industrial Hygienist, was employed in 1973 to direct the project.

The third entity, The National Farm Medicine Center (NFMC), was formed in 1981 as a program of the Marshfield Medical Research Foundation (Hessel & Strack, 1999). The MMRF is an arm of the Marshfield Clinic, which began in 1916 in Marshfield, Wisconsin. Located in a major agricultural area of the state, the clinic had always served the medical needs of farmers. But it wasn't until 1958 and the arrival of Dr. Dean Emanuel, a pulmonary medicine specialist, that the Marshfield Clinic began to focus specifically on farm safety and health. Dr. Emanuel was sparked by the large number of farmers he was diagnosing as having the debilitating disease of Farmer's Lung. This led to pioneering research throughout the 1960s and 1970s into the cause and treatment of Farmer's Lung and other similar respiratory diseases.

In 1979-1980 a task force of Marshfield Clinic specialists met to discuss medical problems they saw in their practice related to agriculture. The result of these deliberations was the NFMC. The initial mission statement describing the organization's interest was to "provide improved health care and well-being for the farmer and farm family and all of agri-business" (NFMC, 1992, p. 2). The NFMC then chose four major functional areas: research, education, clinical services and community services.

The fourth entity, The Bassett Farm Safety and Health Project (BFSHP) at the Mary Imogene Basset Hospital (MIBH), Cooperstown, New York, was formalized in 1983 (Pratt, 1990). The Mary Imogene Basset Hospital is a rural teaching hospital located amidst farm country in rural upstate New York, and was the home base of Dr. John Powers and his pioneering articles on farm injury, although he had moved on to other issues by the early 1960s. There was no other particular focus on agricultural safety and health at MIBH until 1979 and 1980 when Drs. David Pratt and John May, respectively, joined the staff at MIBH. Their specialties were pulmonary medicine. By virtue of a series of research projects on farm respiratory hazards and diseases with area farmers, they felt compelled to personally communicate their findings back to area farmers. At the same time, they became aware of the numerous traumatic injuries suffered by local farmers through personal contacts, hospital records and local media.

While the impetus to form the BFSHP was local, Pratt and May quickly learned that the occupational safety and health problems of local farmers were common to all New York farmers, and indeed, to all of agriculture. A three-year grant from the New York State Department of Labor allowed their project to become statewide. At the same time, Drs. Pratt and May also became active leaders in the national movement for greater public health sector involvement in agricultural safety and health.

Canadians were contributing to the public health sector's understanding of agricultural hazards and motivations for increased involvement in agriculture safety and health during this period as well. Specifically, the University of Saskatchewan's College of Medicine hosted an international symposium on grain dust and health in 1977. The symposium resulted in a collection of papers published in the book, *Occupational Pulmonary Disease: Focus on Grain Dust and Health* (Dosman & Cotton, 1980). The conference organizers, Drs. James Dosman and David J. Cotton, were both pulmonary medicine experts. As suggested by the symposium title, and as borne out by reviewing the published proceedings, the focus of this meeting was on occupational respiratory hazards and disease. From the listing of 58 participants, 19 (33%) were from the United States, and all but two listed an employment affiliation with public health (e.g. departments, schools and colleges of medicine, departments of epidemiology, health, medical centers, NIOSH, etc.).

A second international symposium in 1985, again hosted by the College of Medicine, had a much broader perspective. This conference was titled, "Principles of Health and Safety in Agriculture," and while agricultural respiratory issues still commanded the most attention, a variety of other occupational health hazards and traumatic injury papers were presented (Dosman & Crockcroft, 1989). For example, there were presentations on dematoses, zoonoses, skin cancer, injuries to dairy farmers, machinery accidents to children, injuries from animals, lower back pain of farmers, and stress and mental illness. Of the 183 symposium contributors listed, 92 (50%) were from the United States, and again the vast majority (at least 86%: four individuals listed only a state) were associated with a public health entity.

Other Developments

The media, beginning in the mid-1980s and continuing into at least the early 1990s, played an important role in articulating difficult issues surrounding agricultural safety and health. Several high-profile articles in peer-reviewed journals, and the mass media reinforced the public health sector perspective (e.g. see the "Solomon's Baby" editorial by the Agricultural Safety and Health Committee of the ACGIH, 1980) that no one was minding the store, as it were, when it came to agricultural safety and health. The most critically acclaimed of these was the 1984 Pulitzer Prize winning series of articles "Harvest of Harm" by Tom Knudson in the *Des Moines Register* (September 16-23). Articles also ran in the *New York Times* (Wilkerson, 1988), *The Wall Street Journal* (Ingersoll, 1989), *USA Today* (Keen and Wallmer, 1989), *The Boston Globe* (Butterfield, 1990), and in two widely read national farm magazines, *Successful Farming* (Tevis & Finck, 1989) and *The Farm Journal* (Gaige, 1988), which shed fresh public light on the many problems surrounding farm safety and health. Sometimes directly and sometimes by inference, these articles were critical of agriculture's effort—or lack thereof—to reduce the number and severity of work-related injury.

Many of the later newspaper and magazine stories about agricultural safety either featured or included the story of Marilyn Adams, who lost her 11-year-old son in a gravity flow wagon entrapment incident in Iowa in 1986 (Adams, 1997). Keith had stayed home from school to help with the corn harvest, and for some reason, entered a gravity flow wagon while it was being unloaded. There were no warnings on the wagon about the type of danger that claimed Keith's life. Fueled by her daughter Kelley, a high-school senior, who was determined to make a presentation to the state FFA organization about Keith's incident, Marilyn eventually became a highly visible activist for children and farm safety. A year after the death of her son, Marilyn founded "Farm Safety 4 Just Kids," a nonprofit, grassroots, education-oriented organization whose mission has since grown from education about the dangers of flowing grain to the prevention of all types of farm-related childhood injuries, health risks and fatalities. Farm Safety 4 Just Kids (FS4JK) works to increase public and farm family awareness and understanding of farm hazards, provides educational resources, and sponsors presentations, workshops and community training. Local community chapters of FS4JK present school assembly programs, mount promotional campaigns, and conduct family seminars and farm safety day camps.

One other important reason the public health sector became more interested in agricultural safety and health stemmed from the National Institute for Occupational Safety and Health's (NIOSH) new fatal injury surveillance database. Development of the new database, named the *National Traumatic Occupational Fatalities (NTOF)*, began in 1985 by the Division of Safety Research. By 1987, agriculture as an industry was identified as having among the highest rates of fatal injury of all industries in the nation (CDC, 1987). Although this was not strictly new information (the NSC had been identifying agriculture as a hazardous industry for several decades), the NTOF database represented a census-based approach to occupational fatality surveillance and provided more reliable statistical justification for public health sector concern about the hazards and risks in agriculture, and for involvement in efforts to reduce these risks. The first presentation of the NTOF database results identifying farming and agriculture as having elevated rates of fatal injury was in 1988 at the "Agricultural Occupational and Environmental Health: Policy Strategies for the Future" conference (Myers, 1990).

Finally, farm safety program leadership by agricultural groups declined throughout the 1980s. Most notably, state extension farm safety specialist and state farm bureau safety leader positions were often not filled upon the retirement of the current specialist or leader. National leadership for farm safety similarly declined. For example, the full-time USDA Extension Safety program leader position was replaced with a half-time appointment, and the NSC professional staff for agricultural safety and health was reduced from three positions to one. Those outside of agriculture often viewed these vacant, reduced and dropped positions as a sign of a lack of commitment by traditional agricultural groups to farm safety and health.

Coalescing Into Action

A significant event in 1988 was the "Agricultural Occupational and Environmental Health: Policy Strategies for the Future" conference, hosted by the University of Iowa's Institute of Agricultural Medicine and Occupational Health (Donham, 1990). A four-part conference held during September and including the results of a survey of farmer perspectives on their knowledge and concerns about safety and injury prevention, resulted in the highly acclaimed report, *Agriculture At Risk: A Report to the Nation* (Merchant et al., 1989). More than 170 professionals and private citizens were involved in some way in the conference and the development of the subsequent report. It represented the most significant effort to date to bring together traditional agricultural safety practitioners, public health professionals and political leaders for the cause of farm safety and health. A new committee, the National Committee for Agricultural Safety and Health (N-CASH), was formed to carry forward the goals and policy recommendations that came out of the conference. The committee was comprised primarily of members of the conference policy working group, and included 29 people.

A major goal of the conference was to develop agricultural safety and health policy recommendations in four broad areas:

1. Occupational health and safety services,
2. Education,
3. Health evaluation and research, and
4. Hazard evaluation, prevention and health and safety promotion.

Numerous policy statements were developed within these areas. Environmental health was a second major focus area of the conference. This focus centered primarily on the adverse effects of nitrates and pesticides on groundwater supplies, and several policy statements were articulated. Specifically regarding the agricultural safety and health policy recommendations, the report noted, "All of the initiatives are dependant on greater involvement of the private sector, the government, farm constituency groups, and academia ..." (Merchant et al., p. 29). The report was widely distributed to the media, congress, farm groups, agribusiness, and safety and health professionals, and laid the foundation for the National Institute for Occupational Safety and Health's (NIOSH) entry into agricultural safety and health with a major initiative. It was clearly a conference of unprecedented scope and ambition, but a mere harbinger of what was to come.

The National Initiative in Agricultural Safety and Health

By 1990, several forces had surfaced and merged to bring unmatched and largely negative attention to agricultural safety and health. The NIOSH NTOF system had confirmed agriculture as a hazardous industry and farming as a hazardous occupation, the media had painted a very dark picture of farm safety and health for the public, and a new grass-roots organization had been born out of the tragedy of yet another farm work incident involving a child. These developments, along with a new national report with the provocative title of *Agriculture At Risk: A Report to the Nation*, were finally enough to, once again, attract the attention of the U.S. Congress. With the support of Tom Harkin (D-Iowa), Chair of the Senate Appropriations Committee, the 1989 Congress established a new National Initiative in Agricultural Safety and Health under the direction of the Centers of Disease Control (CDC) (Public Law 101-517, Senate Committee Report). Included in the language of the law was an appropriation of $11.5 million for FY 1990. Funds for the National Initiative were increased to 19.5 million for FY 1991, and to $23.5 million for FY 1992. The funds were decreased to approximately $22 million for FY 1993 and to $21 million for FY 1994, and have been continued at approximately this level ever since.

One part of this federal initiative directed NIOSH to convene a Surgeon General's Conference on Agricultural Safety and Health. The purpose of the conference, held in Des Moines, Iowa, April 30 – May 3, 1991, was to raise consciousness, build coalitions, disseminate information, and encourage action. It brought together more than 400 professionals from public health, medicine, labor, education, legal services, media, and agriculture. The broad areas of surveillance, research, and intervention were addressed in plenary sessions, concurrent technical presentations, and poster presentations. Six unifying principles emerged (Myers et al., 1992): (1) continuous improvement through prevention, (2) recognize the needs of the population at risk, (3) surveillance to measure improvements, (4) research to find root causes, (5) respect people while controlling the problem, and (6) understand "the system" in order to control the problems.

The passage of P.L. 101-517 in 1989 represented the most important public-policy shift concerning occupational agriculture safety and health since the OSHA reform legislation in 1976. To review briefly, agricultural safety and health public policy has historically ranged from exclusionary to one of benevolence. For instance: (a) agriculture was originally excluded from worker's compensation when these laws were passed in the first part of the 20th century, (b) 14- and 15-year-old youths can be exempted from the 1966 Hazardous Occupations Order for Agriculture, and (c) in 1976 OSHA regulations became unenforceable on farms with 10 or fewer hired employees. The major exception to an exclusionary approach to occupational agriculture safety and health public policy was in 1974 when Congress passed the appropriation for an extension safety specialist position in every state. See Kelsey (1991, 1994) and Murphy (1992) for a more detailed discussion on these topics. The point here is that P.L. 101-517—and the specific funding that accompanied it—clearly signaled that occupational agriculture safety and health was now officially considered by federal public-policy makers to be a public health concern to be addressed by the traditional public health approach of surveillance, research and intervention. Congress later emphasized this point when it provided funding for NIOSH to develop a National Childhood Agricultural Injury Prevention Initiative in 1996 (NIOSH, 1999b).

One result of this public policy shift has been to move the agenda for agricultural injury prevention away from traditional agriculturalists and toward the public health sector. This is not saying that traditional agriculturalists, such as cooperative extension specialists, farm bureau safety directors, machinery manufacturer representatives, state departments of agriculture, voluntary safety committees, and others are not still deeply and intimately involved in agenda setting and in efforts to improve agricultural safety and health, because in fact they are. And this involvement is occurring both independently of the public health sector and collaboratively with it.

For example, USDA continues to support farm safety extension programs in all 50 States and Puerto Rico and expanded this support to include the Virgin Islands and Pacific Territory Institutions in 1998. In addition to this, the USDA and the National Easter Seal Society established the National AgrAbility Project in 1991; NSC created the National Education Center for Agricultural Safety (NECAS) at Peosta, Iowa in 1995; and in 2000, the USDA Cooperative State Research, Education and Extension System (CSREES) formed a new committee, NCR-197 Agriculture Safety and Health Research, to prioritize and guide land-grant university research in agricultural safety and health. CSREES received a new appropriation in the federal fiscal year 2001 to implement a competitive grant program for youth farm safety education and certification, and a special research grant to North Carolina for farm injuries and illness. Appropriations were also increased for the formula funds for the farm safety extension programs and the AgrAbility Program. The increase will allow the 16 historically black 1890 Land-Grant Institutions and Tuskegee University to compete for AgrAbilty funding. Just one example of collaboration by traditional agriculturalists in a major public health-based initiative was participation of extension safety specialists and representatives of agricultural industry, organizations and farm families in the development of the 1996 national action plan report for children and farm safety (National Committee for Childhood Agricultural Injury Prevention, 1996).

Table 3.1. Extramural programs and projects funded by NIOSH.

Program Area	No. of Projects	Program Period (FY)
Centers for Agricultural Research, Education, Disease and Injury Prevention [a] (CAREDIP [Ag Centers])	2	1990
	2	1991
	2	1992
	1	1995
	1	1996
	1	1998
	9 total	
Farm Family Health and Hazard Surveillance (FFHHS)	6	1990-97
Agricultural Health Promotion Systems (AHPS)	15	1990-93
	3	1991-94
	18 total	
Occupational Health Nurses in Agricultural Communities (OHNAC)	7	1990-95
	1	1990-92
	2	1991-95
	10 total	
Cancer Control Demonstration Projects for Farmers (CCDPF)	5	1990-93
	3	1993-96
	8 total	
Agricultural Safety Promotions System (ASPS)	6	1994-97
Community Partners for Healthy Farming—Surveillance	8	1996-99
	1	1996-98
	9 total	
Community Partners for Healthy Farming—Intervention Research	7	1996-99
Childhood Agricultural Injury Prevention Initiative—National Children's Center for Rural and Agricultural Health and Safety (NCCRAHS)	1	1997-01
Childhood Agricultural Injury Prevention Initiative—Children's Agricultural Safety and Health Research (CASHR)	8	1997 [b]
	1	1997-98
	9	1997-99
	3	1998-01
	10	2000-02
	31 total	
Interventions for Agricultural Safety and Health (IASH)	6	1998-01
Community Partners for Healthy Farming Research—Intervention Research II	6	1999-02

[a] Ag Centers were funded in different years and have had continuous funding through a competitive renewal process once started.

[b] Limited funding for one year to support further development of the research plan.

Nevertheless, by virtue of budget and influence, it is fair to say that the public health sector, under the direction of NIOSH, has become the leading force in the United States for agricultural safety and health in the 1990s. NIOSH has implemented the initiative through a series of funded extramural and intramural programs. Table 3.1 identifies major extramural program areas, the number of projects funded under each area, and program periods. In addition to these programs, NIOSH has provided extramural funding to support agricultural safety and health initiatives to several of its regional Education and Research Centers (ERC), and has funded research projects (e.g., retrofitting ROPS) through their ongoing R01 research grants program.

Program Components

The first five program areas listed in Table 3.1 were specifically identified as extramural funding priorities in the originating legislation (P.L. 101-517). Of these five, only the Ag Centers program has received continuous funding. An external program review committee in 1995 recommended that: (a) the Ag Centers and OHNAC programs be continued, (b) the FFHHS program be continued only long enough for states to complete existing projects, (c) the AHPS program be folded into other components of the National Initiative, and (d) the Cancer Control program be transferred to other NIOSH initiatives (Kennedy, 1995). The reasoning for these recommendations, and a fuller explanation of them, can be found in the Kennedy report.

A few observations about certain programs of the National Initiative are relevant to this review and are briefly identified here. Ag Centers are located primarily at public health-based entities, such as medical centers and schools and colleges of public health, and are headed primarily by public and environmental health scientists and physicians. The institutions and the principal investigators that are awarded Ag Center cooperative agreements have a common characteristic—significant research accomplishments. Consistent with their research and epidemiological backgrounds, Ag Center activities have focused a large majority of their resources on identifying, measuring, and evaluating farm hazards, risks and exposures (research). The 2001 Requests for Proposals (RFA) for Ag Center funding for FY 2001-05 reinforced an emphasis on research. The RFA required budgeting of direct costs in four core areas: administration, maximum 15%; research, minimum 40%; education and outreach, minimum 10%; and prevention/intervention, minimum 10% (CDC, 2001). A research focus is, of course, consistent with the purpose and mission of NIOSH.

The Farm Family Health and Hazard Surveillance (FFHHS) projects were, as noted in the program review (Kennedy, 1995), about a health and hazard baseline status *survey* rather than surveillance, as the program title suggested. Collectively, the six projects obtained fairly detailed data concerning the health status of farm operators and farm workers, and on workplace hazards. Within the two target areas—health status and workplace hazards—each project decided what variables were important to collect information on, and how they would collect the data. A major goal of the FFHHS program was to provide a national cross-sectional picture of the health status of farm workers and of workplace hazards on farms. While no such report has been assembled, there are several articles in the literature that report on segments of individual projects (e.g., see Browning et al., 1998; Pederson et al., 1999; Zwerling et al., 1998), and comprehensive reviews of the agricultural injury and illness literature cite articles containing FFHHS data (e.g., McCurdy and Carroll, 2000).

The original objective of the Agricultural Health Promotions Systems (AHPS) program was to establish agricultural injury and disease prevention programs at each of the land-grant universities, utilizing the existing network of the cooperative extension service during a three-year period of time. Of the original extramural programs, this was the program area most clearly focused on the intervention component of the National Initiative. Major goals of the AHPS program were to: (a) support existing and new safety education efforts of cooperative extension safety specialists, and (b) take advantage of the access that the cooperative extension system provided to farms, farm operators, and farm families for National Initiative objectives and components. Eighteen projects were funded under this program. Projects were fully funded for three years and received partial funding for a fourth. The lack of individual project impacts, or of outcome evaluation results, appears to have resulted in a shifting of the objectives of this program component in 1994. Rather than broadly support a wide variety of injury and illness prevention programs and activities of the cooperative extension that were difficult to evaluate, the focus shifted to more specific intervention projects that could be evaluated. The new program, Agricultural Safety Promotion Systems (ASPS), resulted in six funded projects but was discontinued in 1998.

The goal of the Occupational Health Nurses in Agricultural Communities (OHNAC) program was to provide surveillance of occupational agriculture disease and injury by placing an occupational health nurse in an agricultural community. OHNAC projects typically incorporated field investigations of targeted occupational injury and illness cases that, in turn, provided data and information for education, outreach and intervention programs. OHNAC nurses often worked collaboratively with extension specialists, county extension agents, county farm organizations and other state and community groups. Operationally, this program was the most similar to the AHPS program and its successor, the ASPS program. After the initial funding cycle, the OHNAC program was renamed "Community Partners for Healthy Farming" and divided into surveillance and intervention programs. A second round of intervention research projects were funded in FY 1999.

The Interventions for Agricultural Safety and Health (IASH) program was created by NIOSH largely in response to informal agricultural initiative program reviews. These reviews suggested a need for more innovative studies and demonstration projects that could test the efficacy of specific interventions, including control technologies, educational programs, health promotion activities, and community-based initiatives. In addition to informal assessments, this program reflected a gradual move by all federal agencies for development of evaluation strategies mandated by the Federal Government Performance and Results Act of 1993. One goal of the IASH research program was to have new products, programs and methods for agricultural safety and health that could be applied regionally or nationally. Six projects were funded in 1998: it is undetermined whether or not the goal of new products, etc., was fulfilled. This program was funded for one cycle of three years and discontinued.

In 1997, with congressionally targeted funds, NIOSH began its Childhood Agricultural Injury Prevention (CAIP) initiative. This program grew out of recommendations contained in the 1996 report, *Children and Agriculture: Opportunities for Safety and Health, A National Action Plan* (NCCAIP, 1996). A national resource center, the National Children's Center for Rural and Agricultural Health and Safety (NCCRAHS), and numerous extramural projects have been funded under this initiative. Many of the early projects served as pilot and developmental projects for the round of proposals funded in FY 2000. One significant project that has been completed was the development of guidelines for children's agricultural tasks. Led by the NCCRAHS, the North American Guidelines for Children's Agriculture Tasks (NAGCAT) were published in 1999 and have been distributed nationally and internationally.

NIOSH has also had several intramural projects in agricultural safety and health, however programs, projects and funding periods are not as easily identified as are extramural programs. There are many reasons for this. First, NIOSH already had several projects under way when the agricultural initiative began in 1990 (e.g., "Occupational Respiratory Disease Surveillance in Agriculture" and "Microbial Exposures in Agriculture"). Second, there are projects of which agriculture is only a portion of the total effort. For example, the organization's Fatality Assessment and Control Evaluation (FACE) and Sentinel Event Notification System for Occupational Risk (SENSOR) projects have included identification of agricultural injury cases. A third reason is that many projects are ongoing. For example, data analyses and publication activities of injury survey projects can continue indefinitely or until there is no more useful information to be obtained from the data (which may be several years). A fourth reason is that some programs, such as the Traumatic Injury Surveillance of Farmers (TISF) survey can be in a pilot-testing phase for a number of years before being budgeted for full-scale implementation. Finally, NIOSH uses its general budget to support base programs, such as the NTOF surveillance program, that have provided data pertinent to the agricultural injury prevention initiative.

Nevertheless, several specific examples of agricultural safety and health intramural projects have been identified since 1990 and are listed in Table 3.2. Most intramural projects involve either injury surveillance with particular populations, research projects to identify, measure and evaluate hazards and hazard exposures, or control technologies. The sample of projects listed below is by broad category with project title and project emphasis. A more complete listing of projects from 1990-1997 is available in *Agriculture: NIOSH Research Projects,* DHHS (NIOSH) Publication No. 97-124 (NIOSH, 1997). The identification of more recent projects is available from the Senior Scientist for Agricultural Research at NIOSH in Morgantown, West Virginia.

Table 3.2. Examples of NIOSH intramural research projects for agricultural safety and health.

Category	Project	Project Emphasis
Surveillance	Traumatic Injury Surveillance of Farmers (TISF)	Survey for work-related injury of all farming operations in the United States.
	National Agricultural Workers Survey Health Supplement (NAWS-Health Supp.)	Survey for nonfatal injury to youthful (<20 yrs. old) farm workers and to children of adult farm workers.
	Childhood Agricultural Injury Survey (CAIS)	Survey for nonfatal injury to children of farm operators and to children visiting farms.
	National Electronic Injury Surveillance System (NEISS) Childhood Farm Injury Follow-back Survey	Surveillance for nonfatal farm-related injury identified through the National Electronic Surveillance System (NEISS).
Hazard and Exposure Assessment	Anthropometry of Construction and Agriculture Populations	Establish a human-form construction and agriculture worker database to evaluate the interaction of workers with their tasks, tools, machines, vehicles and personal protective equipment.
	Agricultural Dusts: Field-Based Evaluation of Exposures and Respiratory Illness	Identify relationships between organic dust exposure and respiratory disease as a foundation for providing control recommendations for prevention.
	Behavioral Health Risks in Farm Families	Examine neurobehavioral disorders and stress-related injury among farm families.
	Herbicide Exposure Assessment Among Custom Applicators	Assess the exposure of custom applicators to seven herbicides.
Hazard Control	Development of Automatic ROPS	Develop an automatically deployable ROPS that will meet industry design standards.
	Control Technology for NIOSH Surveillance Activities	Identify, evaluate and/or develop engineering controls to minimize the incidence of injury/disease identified by surveillance.

The National Initiative Today

From this brief review, and by considering the data presented in Chapter 1 and the earlier discussion in this chapter, a few observations about the National Initiative can be made. First, the National Initiative is directing an unprecedented level of funding and resource support toward agricultural safety and health. For example, approximately $20 million has been directed toward agricultural injury prevention per year for the past 11 years. As a comparison, even if USDA and land-grant universities were funding full-time extension safety specialists in each state at a hypothetical cost of $100,000 per year (salary plus fringe benefits), this would total only $5 million per year. The resource expertise, as evidenced by membership affiliations in NIFS, Inc., which the National Initiative has brought to bear upon agricultural safety and health, is also unprecedented. The National Initiative has also raised the importance (value) of farm safety and health as a program area at land-grant universities by virtue of paying institutional indirect costs on extramural projects. This may amount to thousands of dollars per year on large, multi-year projects. Most USDA cooperative extension funds that support farm safety and health programs do not pay indirect costs. It is not difficult to understand why university administrators value program areas that pay institutional indirect costs over programs that do not.

A second observation has to do with the traditional agricultural injury prevention approach and cooperative extension. The traditional agricultural approach has relied heavily on general safety education, and has been criticized by the public health sector. The initiating legislation for the National Initiative directed the establishment of agricultural injury and disease prevention and health promotion programs at each of the land-grant universities. To state the obvious: the general education approach has emanated largely from cooperative extension programs at land-grant universities.

The support for cooperative extension programs never materialized as envisioned by the originating legislation. As shown in Table 3.1, the prevention promotion component (AHPS) provided funding to 18 state cooperative extension programs for a maximum of four years, and then was discontinued. A justification for not continuing the portion of the component that was implemented was a lack of program evaluation (Kennedy, 1995). The issue of program evaluation will be addressed more fully in Chapter 4. For now, the important point is that while past methods and approaches have been criticized, newer methods and approaches have not yet shown themselves to have a significant impact. One possible reason for this may be that newer methods and approaches have not yet had enough time to become sufficiently widespread to have made a difference. Another possible reason is that newer approaches and methods may not be all that different from past approaches. As the National Initiative has strived to "recognize the needs of the population at risk" and to "respect people while controlling the problem"—two of the unifying principles emerging from the 1991 Surgeon General's conference—it may be that truly different approaches to agricultural injury prevention are exceptionally difficult to come by. This perspective is discussed more fully in the Nature of Farm Work section in Chapter 4.

A third observation is that National Initiative dollars have become concentrated in fewer program components. For example, of the five original program components identified in the initiating legislation, only two components appear to be ongoing in FY 2001. These are the Ag Center and the Community Partners components (the children's initiative is separately funded). The initiating legislation conceived of Ag Centers (hereafter called centers) as research centers that would fulfill the following missions:

1. Develop model programs for prevention of illness,
2. Develop model educational programs on agricultural safety and health,
3. Evaluate injury and disease prevention programs implemented by agricultural extension programs,
4. Conduct applied research and evaluations of engineering and ergonomic controls and procedures, and
5. Provide consultation to a variety of others interested in agricultural safety and health.

The number of centers has increased from two in 1990 to nine in 1998. The recent RFA for centers indicates funding of seven to nine of them for up to $1,000,000 each per year over a maximum of five years. Clearly, the center concept is one that NIOSH is supporting.

Possible results of concentrating funding into fewer program components include: (a) more dollars become tied up in institutional indirect costs resulting in fewer dollars available for direct program costs, or (b) less collaboration between the centers and the states within their region. Collaboration between states and a center is most often achieved by a center's funding of research and intervention projects through subcontracts proposed by collaborators at land-grant universities, state government, or other nonprofit entities. Such collaboration is one strength of the centers' concept, and at least one external evaluation of the National Initiative encouraged more of it (Kennedy, 1995).

However, the more subcontracts a center enters into as a means of collaboration, the greater percent of fixed total dollars that go toward institutional indirect costs. The rules that govern institutional indirect costs involving federal dollars and subcontracts (US OMB, 2000) limits institutional indirect costs to the primary institution (e.g., a center) to the first $25,000 per subcontract, including multi-year subcontracts. Thus, for a large (e.g., $200,000), multi-year subcontract, institutional indirect costs do not become a significant issue. But if an Ag Center wishes to subcontract this same dollar amount with three or four different cooperators, or wishes to fund multiple projects with a single cooperator, total institutional indirect costs become more significant because each individual project will be for fewer total dollars, and the first $25,000 becomes a larger percent of each subcontracted project. For example, with a subcontract of $200,000 and an indirect rate of 40% (a representative figure for many universities; some nonacademic institutions may have rates considerably lower) limited to the first $25,000, the indirect charge applies to only 12.5% of the total grant ($25K of $200K). However, if the same $200,000 dollars are divided into four $50,000 subcontracts, the indirect charge applies to 50% of each subcontract, leaving far fewer dollars to support direct program costs. In each of the scenarios above, the institution receiving the subcontract further reduces project dollars with their own institutional indirect cost charge on direct program costs, and this occurs without any limits (e.g., the first $25K).

One way to avoid so many dollars being lost to indirect costs would be for NIOSH to bypass centers and fund more projects directly with other institutions, negating National Initiative dollars being reduced by two separate institutions. But this idea may hinder the regional and collaborative approach to agricultural injury prevention that is afforded by the centers concept. A geographical-based region (or multi-state) approach has considerable intuitive appeal because many types of farming, work hazards and risks, populations exposed, etc., easily cross state borders while other important factors, such as weather, growing seasons, terrain, and economic, social and political climates, often exist regionally. Hence a regional approach helps to maximize scarce resources through collaboration.

A fourth observation is that the centers program has been in a nearly constant state of growth from the beginning of the National Initiative, with different centers funded across an eight-year period. Issues the centers have addressed are different start-up years, levels of funding, number of states within a region, and levels and types of production agriculture and work injury problems. As more recent centers were established (e.g., the Southwest Center in 1995, the Pacific Northwest Center in 1996, and the Deep-South Center in 1998), some existing ones were able to refocus on a smaller geographic area, while the new ones were able to focus on a smaller geographic area from the beginning. All these differences and changes have resulted in—as would be expected—considerable variation in program priorities, activities, and reporting by centers over the eight years of their existence.

All these differences make it difficult to comprehensively measure the centers component as a conceptual approach to agricultural injury prevention. A new cycle of funding for centers, beginning in 2001, means that some existing ones may not be refunded, new ones may be funded, or that states associated with a particular center may change. Any of these possibilities add to difficulty in tracking broadly based impacts the centers component may be having on agriculture injury prevention. Nevertheless, the situation is improving as a summary report of fiscal year 2000 activities showed that current centers were developing standardized reporting metrics (NIOSH, forthcoming). This report showed that centers, among other things: (a) were active with a wide variety of state and regional research and outreach projects among 42 different states, (b) have collaborative projects with each other (e.g., the Southeast Center with Deep-South Center and the Southwest Center); and (c) are addressing the needs of both traditional (e.g., Midwest farmers and their families) and less traditional (e.g., Hmong and Lao migrant and seasonal workers) audiences.

Chapter Summary

This chapter has reviewed historical and modern involvement of public health in farm safety and health. Public health has probably had a longer and closer connection to agricultural safety and health than many realize, although it did not become nationally organized until the late 1980s. In the beginning, physicians lead public health participation in the farm safety movement by summarizing farm injury incidents and their causes. As a wider variety of public health-based professionals became interested in agricultural safety and health through the 1980s, their primary contribution was in the area of descriptive epidemiology, most notably as it related to occupational health hazards, risks, and exposures. As physicians, public health epidemiologists, researchers and clinicians became involved in agricultural injury prevention, they used primarily educational intervention methods.

Public health-based professionals became increasingly dissatisfied with the general state of affairs in agricultural safety and health in the 1980s. This has lead to an increased presence in the farm safety and health movement by public health, operationalized through the National Initiative in Agricultural Safety and Health. This National Initiative, initiated by Congress, has been implemented by NIOSH through a series of intramural and extramural programs, including a more recent special emphasis on youth safety and health. Two of the five original National Initiative program components remain active, with the centers component receiving the most financial support.

4 Fundamental Challenges to Agricultural Safety and Health

This chapter reviews historical and contemporary impediments to reducing occupational agriculture injury. The impediments identified here permeate and negatively impact progress in agricultural safety and health in the United States. The discussion is organized around three main themes: the nature of farm work, program theory and evaluation, and surveillance data issues. These themes are related only in that they represent long-term and difficult problems to overcome. Some of the specific problems identified within these themes were first noted as many as 60 years ago, some problems have emerged only in more recent years, while still others have never been discussed as formally as they will be here. These challenges are interrelated and often feed into each other; perhaps this is why they are not easily addressed or resolved. Nevertheless, these are issues that need a systematic, long-term, ecologic (holistic) approach if major improvements are to be made to significantly improve occupational agriculture safety and health.

The Nature of Farm Work

The first theme area is the nature of farm work. Many things about the nature and practice of farming contribute to its hazardness and the difficulty of reducing its hazards and risk. Clyde Berry (1971), Industrial Hygienist, Institute of Agricultural Medicine, University of Iowa, assembled a list of factors that helped to explain differences between farm and industrial safety and health. Many of these factors and others, outlined in Table 4.1, have been discussed individually in literature, and are readily understood by agriculturalists and experienced farm safety and health professionals and need little elaboration here. However, less often thought about or discussed is the synergistic effect of these factors on agricultural injury prevention. Collectively, these factors have a huge negative impact in changing the status quo of safety and health beliefs and practices on the farm, and pose obstacles for professionals engaged in agricultural safety and health research, education and intervention.

Farm Safety—Risk Paradox

The abundance of hazards and injury risks on farms is clearly documented in literature. The U.S. Secretary of Transportation report (1971), NSC's 31-state *Farm Accident Survey Report* (Hanford et al., 1982), the *Papers and Proceedings of the Surgeon General's Conference on Agricultural Safety and Health* (Myers et al., 1992), and the McCurdy and Carroll (2000) review paper are representative examples of this documentation during the last 30 years (these reports are discussed in more detail later in this chapter). It has also been clearly documented that while farmers understand farm work is potentially dangerous, they seem relatively unconcerned about their risk of injury, particularly in comparison to other, more immediately perceived concerns such as product prices, machinery repairs, work loads, etc., (e.g., see Thu et al., 1990; Elkind, 1993; Elkind & Cody-Salter, 1994; Dewar, 1996; Kidd et al., 1996; Green, 1999; Hwang et al., 2000). Nor do farmers' perceptions of hazards and risks match up well with injury records (e.g., see Rosenblatt & Lasley, 1991; Whitman & Field, 1995; Schwab et al., 1995). Finally, despite parents' concern for their children's safety and health, parents routinely expose their children to the same life-threatening work hazards and risks that they accept (e.g. see Tevis & Finck, 1989; Hawk, et al. 1994; Lee et al., 1997; Darragh et al., 1998). In all, the literature suggests considerable incongruence and large disconnects between farm people's safety knowledge, values, and practices. This incongruence and disconnect may be termed the *farm safety—risk paradox*.

This paradox has been explained in several of the studies cited above, particularly those utilizing focus groups and surveys to ascertain what farmers and farm families think and believe about farm safety (e.g. see Elkind, 1993; Kidd et al., 1996; Lee et al., 1997; Darragh, 1998; & Green, 1999; Murphy, 1992, pp. 212-216). In summary, it helps to identify the safety and health culture of farming by understanding the explanations and reasoning farm people give for safety knowledge, beliefs and practices that seem contradictory to each other. This farm safety and health culture—which can be described as one of risk taking and suppression of concern about serious injury to themselves and loved ones—has its genesis in the combined effect of the factors that make up the nature of farming. Additionally, the culture is reinforced by everyday experiences on the farm. That is, the commonness of the personal experience of interacting with hazards and not being injured, and the earliness at which its starts for many (e.g., in childhood or adolescence), results in an ingrained belief that hazards and injury are as integral to farming as seed and feed.

Table 4.1. Factors that contribute to farming's hazardness and difficulty in reducing hazards and risk.

Factor	Comment
Environmental	
Weather	Farm work must often be completed regardless of weather extremes.
Work sites	Commonly overlaps with residence.
Emergency services	Not readily available, often involves a delayed response due to isolation of work site.
Isolation of work	Co-workers often not in eyesight or hearing distance in case of trouble.
Personal hygiene	Often required and facilitated for in other occupations; largely self-imposed in agriculture.
Environmental hazards (noise, vibration, illumination, dusts, etc.)	Hazards and exposures not monitored or regulated as with most hazardous occupations.
People	
Young workers	Children less than 16 years old—and as young as five—are commonly exposed to and interact with work hazards and environments that are beyond their normal physical, mental and/or emotional ability to safely respond to.
Senior workers	There is no standard retirement age in agriculture. Results in farmers with significant physical limitations (e.g., hearing, vision, balance, coordination, dexterity, strength deficits) and slow reaction times continuing to work in high-risk situations.
Minimum physical limits	Initial physical exams or minimum performance requirements are often required to begin work or to continue work in other hazardous occupations.
Physical exams	Routine medical surveillance not common.
Special provisions for physical or mental conditions (sickness, pregnancy, physical handicap, learning disability, emotional disability, etc.)	Not available or only self-imposed restrictions. Tightly controlled in other hazardous occupations.
Transfers to light duty	Not usually an option in agriculture.
Dispersion of workforce	Difficult to provide health and safety services because of geographic dispersion and mobility of workforce.
Farm operators	Ranges from those with advanced college degrees to those with a high-school education; farm fulltime and work significant hours off the farm; work fulltime off the farm and farm for supplemental income, to farming only as a hobby or lifestyle statement.
Work Activity	
Work hours	60- to 80-hour work weeks common.
Labor and management functions	Usually separate functions in other hazardous occupations but not in farming.
Work pace	Can be highly erratic rather than steady; frequently affected by weather and machinery breakdowns.
Work routine	Can be highly irregular; many tasks are seasonal and may only be done once or twice per season or year.
Specialization	Not normally possible; the phrase, "jack-of-all-trades" frequently applies.
Instruction	Farmers learn their trade by observational and experiential learning.
Holidays and vacations	Days off are normal for most occupations but not in farming.
Labor demands	Frequently make use of any temporarily available labor, including migrant, spouse, children, friends, visitors, new acquaintances, and off-the-street employees. Children growing up on a farm normally viewed as an inexpensive labor source.
Uncertainty	Farming is characterized by an inherently uncertain immediate future. Weather, fast-moving plant and animal disease, broad U.S. economic policy, and unexpected world events can each result in short- and long-term financial hardship for the farmer.
Agriculture production	There is tremendous variability in size and type of production units, and concomitantly, in sophistication of technology used, making it very difficult to accurately characterize modern agriculture.
Social, Economic & Political	
Lifestyle vs. occupation	Farming is commonly viewed as a "way of life" rather than as an occupation. This is why society's approach to occupational hazards and risk—a mostly hands-off approach—is different for agriculture than for other hazardous occupations.
Agrarianism	This is the term applied to agriculture that says farmers are owed a social debt by society because they suffer so that a democratic society can prosper. The U.S. version of this concept traces back to Thomas Jefferson.
Day care	Often not available, practical, or affordable in rural areas. Results in parents baby-sitting infants, toddlers, preschoolers and other children during farm work.
Occupational safety and health legislation	New standards and regulations often exempt production agriculture because of a combination of lack of practicality to farming, lack of ability to enforce the standards or regulations, and the burden on farmers to comply.
Cultural beliefs about farm safety and health	It is a cultural belief that farming is a hazardous and unpredictable occupation. This contributes to the belief by farm workers that little can be done about farm safety and health except to be careful.
Market forces	Farmers don't set their own prices for products produced. They can't add the costs of safety and health to products to recoup costs.
Self-reliance for safety	Farmers primarily rely on their own knowledge and awareness of hazards to work safely, and often accept blame when an injury occurs, especially when they commit an unsafe behavior that directly results in an injury.
Enculturation	Children are taught values, responsibility, good work ethics, decision-making, and about life and death. Strong bonds among children, parents, grandparents, neighbors and communities are developed and nourished from the shared experiences of farming.

Without a doubt, farm safety and health cultural values become a constituent part of the overwhelming cultural ethic in agriculture, which is to get the work done—weather, input costs, product prices, work load, machinery breakdowns, and hazards be damned! Stated in another way, it can be said that farmers interact with the uncertainty, the never-ending work, and the dangers of farming on a daily basis by adopting an attitude of indifference or of resilience that proclaims, "We'll see ourselves through this." Surviving the day without a major injury is a personal testament of knowing how to "handle the hazards" and being "careful enough." There is no reason for them to think tomorrow will be any different. This reasoning, of course, is seemingly correct, day after day, sometimes year after year, for most farmers because they may have farmed for years without suffering a serious farm-related injury themselves, nor had one happen to someone else on their farm.

The *farm safety—risk paradox* is as problematic for agricultural safety and health professionals as it is for farm families and workers. The paradox suggests that approaches to farm safety research, education and intervention must consider the interconnectedness among the many factors of the nature of farming, and how these factors influence prevailing cultural beliefs and practices. In turn, this implies that approaches must embrace fundamental social, political and economic realities shaping farming's common culture. Many of the presentations made during the 1991 Surgeon General's Conference on Agricultural Safety and Health (Myers et al., 1992) recognized that an ecologic perspective of agricultural safety and health was important, and Scharf et al. (1998) reviewed projects that illustrate how this perspective is beginning to take place in field research and demonstration projects. The 1997 Tractor Risk Abatement and Control Policy Conference at the University of Iowa provides further evidence that a holistic approach to a specific farm safety issue can be formulated (Donham et al., 1997). Many questions regarding the long-term sustainability and effectiveness of these projects are as yet unanswered. More importantly, such projects must become the norm rather than the exception.

Program Theory and Evaluation

The second theme area is program theory and evaluation. Exacerbating the impact of the *farm safety—risk paradox* is that agricultural safety and health professionals have historically lacked well-articulated, research-based theory or models to guide injury prevention and educational programs. Collegiate dictionaries define "theory," in part, as: "a coherent group of general propositions used as principles of explanation for a class of phenomena," and "a proposed explanation whose status is still conjecture." From an occupational safety and health perspective, program theory refers to the set of beliefs and assumptions that undergird injury prevention programs (Weiss, 1997).

In practice, injury prevention program theory is a construction of plausible and sensible ideas of how an injury prevention program is supposed to work (Bickman, 1987). Initially, these plausible and sensible ideas come from the insight and practical experiences of those designing the programs. To move beyond intuition and hunches, though, requires the development of research hypotheses that can be tested in laboratories and field settings. Hypothesis testing serves as a vehicle for moving plausible and sensible ideas about how a program works from conjecture and guessing to more objective, scientific-based statements of cause and effect. Understanding as precisely as possible program cause and effect is, of course, a main purpose of program evaluation. It is in this way that program theory and program evaluation are directly linked and feed into one another. This linkage is increasingly referred to as theory-driven (Chen, 1994) or theory-based evaluation (Chen, 1994; Weiss, 1997).

Contemporary program evaluation theory has a focus on program outcomes or impacts along with an assessment of program characteristics to better understand what actually goes on in a program (Conrad & Miller, 1987). Program evaluators have recognized for a long time that experimental programs in field settings are typically neither implemented as planned nor delivered to participants in a fixed, standard manner (Finney & Moos, 1989). Hence the purpose of contemporary program evaluation theory is to direct attention to critical program activities, the intended outcomes, and presumed relationships between those activities and the intended outcomes (Lipsey & Corday, 2000).

Past and Present

Injury prevention in agriculture is typically education oriented (even an engineering solution, e.g., ROPS for tractors, relies strongly on educational programs to convince farm operators and workers to use the solution). In today's parlance, one speaks of this as an "intervention." A clear statement of theory helps to identify the assumptions about human behavior that underlie a particular intervention (Scheirer, 1994). The concept of using well-specified theory to drive or justify intervention programming is just beginning in agricultural safety and health. It is occurring in the context of behavioral theories and models that attempt to explain and predict voluntary farm safety behaviors. This topic, behavioral theory and models, will be discussed more fully in the next chapter. For now, the important point is that there is little historical evidence in the agricultural safety and health literature that either program theory or program evaluation has ever specifically driven injury prevention interventions.

DeRoo and Rautiainen (2000), in their systematic review of farm safety interventions, found only 25 articles on farm injury prevention interventions, even though they accepted intervention studies that did "not meet the usual methodological criteria of systematic reviews" (p. 52). Of the 25 studies cited, nine were included even though evaluations of the interventions were still to be completed, or were not planned. This point is made to underscore the general lack of program evaluation in agricultural safety and health, although this is not a new observation (Butler, 1974; Merchant et al., 1989). Regardless of theory, there is a developed body of literature on intervention evaluation that can be discussed independent of theory. This literature has direct relevance to agricultural safety and health, and is the topic discussed in the remainder of this section.

Scientific program evaluation became firmly established in this country in the 1930s and experienced a boom period after World War II (Rossi et al., 1998). In the early 1970s, intervention evaluation research emerged as a distinct specialty in the social sciences and is today commonly applied to education, public health, and public policy issues. Recent state-of-the-art reviews of program evaluation (e.g. see Weiss, 1997; Lipsey & Corday, 2000) suggest that it is a field of expertise still marked by many fundamental questions regarding theory, methods and applications. For example, some perspectives view theory as the starting point for evaluations, while others view it as an end point (Lipsey & Corday, 2000).

There is no particular secret to understanding why the application of scientific program evaluation by traditional agriculturalists has not been a major component of agricultural injury prevention programs. Educational background, job expectations, and tradition all play prominent roles. First, and as documented in Chapter 2, the academic backgrounds of most of the early cooperative extension farm safety leaders were in general agriculture or agricultural engineering. Scientific program evaluation, at any level of sophistication, would not have been a major part of their academic training, if the topic was addressed at all. Nor would scientific program evaluation have been a part of the education or on-the-job training of nonacademic leaders, such as representatives from machinery manufacturers, agricultural services, and farm organizations. Thus, the skills needed to incorporate quality program or intervention evaluation were absent.

Literature shows little evidence suggesting that the cooperative extension system has ever expected sophisticated program evaluation from its safety specialists, even though evaluation has been recognized as a valuable component of agricultural safety and health programs throughout its history (e.g., see Mohler, 1937; Gilden, undated; Greenwood, 1983). This is important because the Cooperative Extension Service assumed a primary leadership role in the development and implementation of farm safety education programs from the very beginning of organized farm safety education (Wood, 1963), particularly at the state level (Erickson, 1940; Hoffman, 1952; Wardle, 1958).

Historically, the emphasis in extension safety education has been on the *doing* of safety education, rather than the evaluation of it. Consequently, there has been no tradition upon which to build strong, scientific, safety program evaluation. Green et al.'s (1980) warning to health educators about heath education programs seems very appropriate for extension safety educators. Green et al. said practitioners could fall into a trap when they become so involved with the "doing" of educational programs that they lost sight of why the program was offered in the first place. Of course, the trap is to continue an educational program with no evidence of its effectiveness. Given the difficulty of performing scientific program evaluation even when educators want to (Dannenberg & Fowler, 1998), a significant change in the status quo of extension safety education program evaluation will at least be difficult because of a lack of resources for program evaluation and competing program pressures (e.g., obtaining and maintaining organizational and financial support, development and implementation of multiple programs across a given state, being responsive to clientele groups for technical information, etc.).

The public health sector also does not have a good historical record of scientific program evaluation in agricultural safety and health. The physicians and environmental health specialists that have always led public health sector involvement in agricultural safety and health—in keeping with their training and professional practice expectations—have concentrated on the research side of agricultural safety and health, not its educational program side. The science of epidemiology can offer insights into describing injury and illness (descriptive epidemiology) and cause and effect relationships (analytical epidemiology), but it does not necessarily inform on the ways and means of injury and illness prevention in real work settings. For example, documenting that tractor overturns result in the highest rate of fatal injury to farmers may suggest ROPS (as the intervention) on a tractor is critical, but it does not inform us on how to convince a farmer to install a ROPS on a tractor without one. Similarly, differentiating among the various types of mold spores that cause respiratory disease does not inform us on how to actually convince farmers to institute environmental dust control technology or to wear respiratory protection. Epidemiological investigations, of course, are not meant to provide details on behavioral motivations.

The research literature suggests that scientific safety and health intervention evaluation is not yet a part of the culture of agricultural safety and health professionals, whether they are agriculturally-based or public health-based. The authors of two systematic reviews of the farm safety and health injury prevention literature noted major methodological flaws with the majority of studies of agricultural safety and health interventions that they reviewed. For example DeRoo and Rautiainen (2000) noted the lack of randomization of study subjects and control groups, no baseline measurements for some interventions, same-day pre- and post-tests, a lack of objective measures of outcome, and inadequate data for process evaluations. Keifer (2000), in his and a colleague's review of intervention effectiveness in reducing pesticide overexposure and poisonings, similarly found a lack of randomness, control groups, objective measures of outcome, and small sample sizes. Such weaknesses, while typical of many occupational safety and health studies, seriously limit the quality of intervention evaluation research.

Toward Change

These criticisms notwithstanding, the national cooperative extension system has had a widely recognized program evaluation model—Bennett's Hierarchy—in place since the mid-1970s (Bennett, 1975; 1979). It was recently expanded and updated to the Targeting Outcomes of Programs (TOP) model (Rockwell & Bennett, 2001). Public health has also had health program evaluation models, e.g., the PRECEED model, since at least the mid-1970s (Green et al., 1980). This too, has been expanded and updated as the PRECEDE—PROCEED model (Green & Kreuter, 1991). In 1992, Teutsch proposed "A Framework for Assessing the Effectiveness of Disease and Injury Prevention" for the public health service, and more recently, a CDC Evaluation Working Group developed a "Framework for Program Evaluation in Public Health" (CDC, 1999).

A few recent farm safety intervention projects recently have shown positive results. For instance, Mandel et al. (2000) found increased use of gloves and other protective equipment in the intervention group compared to the control group; Legault and Murphy (2000) found that intervention group farmers reduced hazards using the *Agricultural Safety and Health Best Management Practices* (ASHBMP) manual more than a control group of farmers; and Carrabba et al. (2000) found that 4-H tractor program drivers were observed to operate tractors in a safer and more skillful manner than tractor program drivers not involved in 4-H. Additionally, Landsittal et al. (2001) found that three different educational intervention groups within a single study reduced farm safety hazards more than the control group. Clearly, program evaluation in agricultural safety and health is moving in the right direction.

Lipsey (1996) has noted that since program evaluation emerged in the 1970s, its history has been one of continuing revision and elaboration. As an example, Glasgow et al. (1999) recently introduced "The RE-AIM Framework" for evaluating the public health impact of health promotion interventions. Their acronym stands for the five functions of reach, efficacy, adoption, implementation, and maintenance. A central tenet of this model is that the ultimate impact of an intervention is due to its combined effects on the five evaluative dimensions. Their model continues the holistic, or ecologic, approach to program evaluation common to public health educators and professional program evaluators (e.g., Green & Kreuter, 1991; Rossi et al., 1999; Wholey et al., 1994). Researchers have noted the lack of comprehensive schemes for evaluating occupational safety and health interventions (Schulte et al., 1996; Shannon et al., 1999; Zwerling et al., 1997). Recent efforts by NIOSH have begun to address this shortcoming.

Intervention Effectiveness Research

A conceptual framework for assessing the effectiveness of occupational safety and health interventions has been proposed by the NIOSH Intervention Effectiveness Research (IER) Implementation Team (Goldenhar et al., 2001). The IER team is one of 21 teams formed to address National Occupational Research Agenda (NORA) priority areas (NIOSH, 1996). The conceptual framework describes an occupational intervention effectiveness research (IER) process (Figure 4.1). The IER process is not a theory-driven evaluation model itself, but theory can be used to guide the specific intervention. This model seems particularly appropriate for guiding agricultural safety and health intervention evaluation because it provides a structured process, in the language of occupational safety and health, to address major inadequacies of past farm safety educational program evaluations. For example, Aherin et al. (1992), Murphy et al. (1996) and DeRoo & Rautiainen (2000) have all noted the lack of experimental evaluation of interventions.

As shown by Figure 4.1, this structured process includes three interdependent and sequential phases that are important to effective safety and health intervention evaluations: development, implementation and effectiveness. Each phase is operationalized by five central tasks that should be carried out in each research phase. A manual, *Guide to Evaluating the Effectiveness of Strategies for Preventing Work Injuries: How to Show Whether a Safety Intervention Really Works*, also developed by the IER team, provides further details on intervention effectiveness research (Robson et al., 2001). These two resources are used to summarize below important aspects of the three phases and five central tasks of the IER conceptual model, particularly as they may relate to issues of farm safety and health interventions.

Development Research Phase

This phase attempts to document how and why certain interventions might be expected to bring about desired changes. Typical questions that might be answered include:

- What changes are needed to enhance the health of the target population?
- What are the best ways to bring about these changes?
- What principles or themes in occupational safety and health and related fields might apply in a particular situation?
- Why have these changes not already been made?
- To what extent does the target audience understand and buy into the need for change?

These types of questions are normally associated with conducting a needs assessment (Rossi et al., 1999). Typical sources of information to answer these questions include surveillance data, hazard surveys, technological innovations, behavioral change theories, results from previous intervention studies and focus-group interviews with the target population.

A significant benefit of answering needs assessment questions and operationalizing the five intervention research tasks in this phase is that it encourages researchers to consider social, political and economic issues up front that might later undermine the intervention if these issues are ignored. For example, an educational intervention may convince a community of farmers that operating a tractor without a ROPS is hazardous and one should be installed on all tractors, but the intervention effort may be of little practical use if the majority of farmers do not have the financial resources to install ROPS on their tractors. The researcher would likely have uncovered this problem by talking with farmers in the targeted community (task 2, developing partnerships) about the proposed intervention (task 3, choose methods or designs). In general, well-designed intervention development research increases the potential that an intervention will be ultimately successful by providing greater detail and understanding of how the intervention might affect the target population. Kidd et al. (1996) used this approach with farmers to develop a link between psychosocial stress and injury, which in turn was used in the development of two specific intervention strategies: Kayle's Difficult Decisions Exercise and the Farm Planning Tool (Kidd et al., 1998).

Implementation Research Phase

Implementation research studies (also called formative evaluation or process evaluation) attempt to systematically document how an intervention was *actually* implemented in comparison to how it was *intended* to be implemented. Typical questions that might be answered during this phase include:

- What are the components of the intervention and how were they provided to the target audience?
- Was the quality of the intervention components adequate (e.g., trainers well qualified, equipment properly calibrated)?
- How many and who in the target audience experienced the intervention?
- Did any target people reject the intervention?
- Was the timing of the intervention correct?
- Did the intervention deviate from expectations, and if so, why?

Documenting answers to these types of process questions often provides understanding of why an intervention either fails or succeeds.

In field research there can be any number of reasons why an intervention fails to produce expected results, many of which may be beyond the control of the researcher. For example, in a mass media campaign a breaking news story of national interest may destroy the timing of public service announcements, or preempt local coverage of an event critical to the campaign. An intervention, for example an educational campaign, to convince farmers to purchase a ROPS, may be undermined by an unexpected USDA

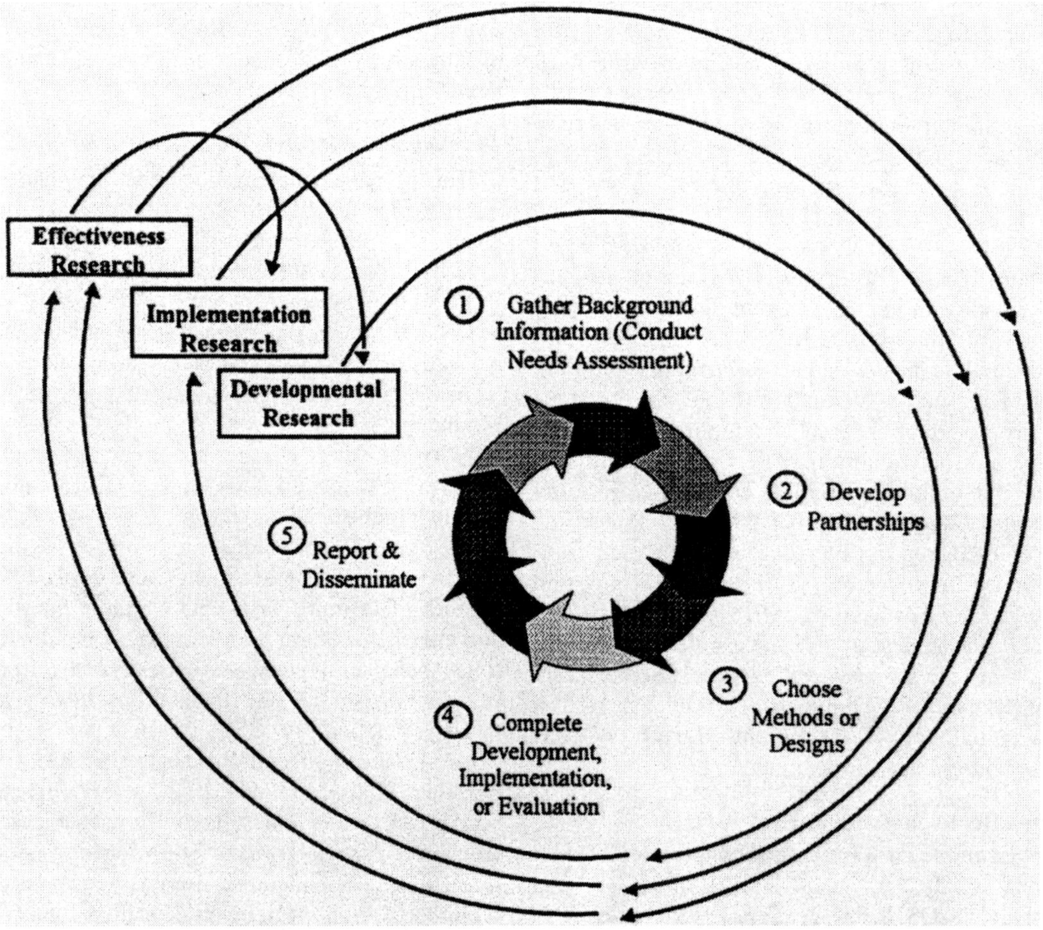

Figure 4.1. Intervention Effectiveness Research (IER) conceptual framework (Goldenhar et al., 2001).

announcement of a commodity surplus that results in a steep decline of market prices. The unexpected drop in prices may inhibit the purchase of ROPS even though farmers were convinced by the educational campaign that a ROPS should be purchased, and had expressed an intention to do so when questioned in a follow-up survey. On the other hand, an increased purchase of ROPS after an educational campaign may be due more to a severe drought in other countries of the world, and subsequent overall market price surge, rather than the intervention itself. In any case, identifying factors that interfere with the conduct of the intervention facilitates a more accurate interpretation of the intervention results. In turn, an accurate interpretation of results is crucial to decisions about replicating the intervention and generalizing the results.

Effectiveness Research Phase

Effectiveness research (also called outcome, impact or summative evaluation) identifies the extent to which an intervention succeeds or fails in the real world. The distinction between "efficacy" and "effectiveness" studies is important to understand. Efficacy refers to studies in controlled settings (for example, in a laboratory, or a field study with a highly motivated group of individuals). Questions that the effectiveness research phase might answer include:

- To what extent does the intervention reduce injury, disability, or disease?
- To what extent does the intervention reduce exposure to hazards?
- What is the effect of the intervention on social and economic consequences of injury or disease?
- How have knowledge, attitudes or behavior changed?

Effectiveness refers to studies under real-world conditions of the target population, where there is much less control of extraneous variables that may impact the intervention. An experimental study with randomized treatment and control groups is the gold standard for determining cause and effect between an intervention and an outcome. However, randomized, control group studies of occupational safety and health can be very difficult to accomplish in real world settings for reasons of ethics; a lack of available comparison groups because they may not be available in the real world of work; time and funding constraints; and other similar factors. Reasonable alternatives to randomized control group studies exist, and often provide a means of reaching a sensible and economic compromise design for determining intervention effectiveness. For example, a variety of techniques exist to strengthen nonexperimental designs, such as using other historical data and multivariable regression, and quasi-experimental designs, such as time series analysis and reverse intervention introduction (Robson et al., 2001).

Many times, when a randomized control study cannot be implemented, collecting both qualitative and quantitative data measures can still enhance experimental value. Qualitative data refers to nonnumerical data such as that gathered by note taking or tape recording interviews, or by holding focus-group discussions or special subject-matter meetings. Quantitative data refers to traditional research measures that are recorded numerically. Advantages of quantitative data include reliability along with collection and analyses efficiencies. However, qualitative data can often add meaning and richness to numeric data. For example, the number and types of hazards that exist on a farm are typically recorded numerically through a structured paper form, for example a self-report hazard survey or audit form. But understanding why so many hazards exist in the face of known remedies is not so easily captured on paper forms. Rather, this information is better captured in the free flowing expressions and explanations of the farm operator.

Intervention Research Tasks

The central intervention research tasks, identified in Figure 4.1, are conceived of as being equally important to all three phases of the evaluation. The tasks are relatively straightforward steps similar to many program planning, implementation and evaluation models (e.g. see CDC, 1999; Rockwell & Bennett, 2001; Seevers et al., 1997; Scheirer, 1994), and need little elaboration here. Nevertheless, from the perspective of the independent nature of farmers and their tolerance of work hazards and risk, paying close attention to these tasks during the developmental phase seems especially important for agriculture safety and health intervention effectiveness. Adhering to the ideas expressed by the five tasks early in the development phase will provide natural opportunities for their consideration in the later phases. Task 2, develop partnerships, can be especially important. For example, in the Pennsylvania Central Region Farm Safety Pilot Project, contact was made with state leaders of three major farm service organizations in the development phase of the project to obtain their input and support for the project (Murphy et al., 1998). Additionally, two meetings were held with county agents representing several counties in the region where the intervention might possibly be implemented. These initial meetings resulted in:

- A co-signed letter from one principal investigator and state farm organization leaders to their respective county leaders encouraging their cooperation with the project. This facilitated project planning meetings with county organizational leaders, which in turn resulted in a co-signed letter from the county agent and the county organization leaders to all farmers in their county urging cooperation with the project.
- At the suggestion of county agents, a switching of specific intervention projects between two counties was done to better match intervention goals and objectives to their own program strengths, and to the farm population in their county.

Not only did these partnerships strengthen final development of the specific intervention projects, but the relationships developed helped to garner support for the implementation phase of the project as well. Because farmers had received a letter of support from their local farm organization, and because agents were better matched with specific intervention projects, the implementation phase was impacted in a positive way.

Professional partnerships are also critical. Areas of expertise that may be needed in an IER project include agricultural sciences, economics, epidemiology, human factors engineering, industrial hygiene, behavioral psychology, program evaluation, and rural sociology. As explained earlier, many agricultural-based educators and safety specialists may not have finely developed intervention evaluation research skills. But many do have an intimate understanding of the work environment and culture of farm operations, and often have credibility with farm families and agricultural leaders. On the other hand, many nonagricultural professionals may have little natural credibility with farm families or agricultural leaders, but bring valuable IER skills to the project. The NIOSH IER team members, in their paper, emphasize the interdependence of the three phases for evaluating an intervention, creating a cycle of continuous improvement. A team of injury prevention professionals that encompasses a wide array of expertise, working together from the developmental phase, can only enhance the effectiveness of agricultural safety and health intervention evaluation research.

Evaluability Assessment

One additional point concerning IER is important to understand. Not all interventions are appropriate for a full IER treatment. An evaluation that moves through all three phases of IER will normally require valuable and limited resources that may be more effectively utilized elsewhere in a safety and health program. One way to explore the feasibility of a full IER effort is to conduct an evaluability assessment. The idea of an evaluability assessment comes from the work of Wholey and colleagues as they analyzed obstacles to evaluating public programs (Rossi et al., 1999). Wholey et al. (1994) has noted that many public and private program evaluations are inhibited by four problems:

1. Lack of agreement on the goals, objectives, side effects, and performance criteria to be used in evaluating the program.
2. Program goals and objectives that are unrealistic given the resources that have been committed to the program and program activities that are under way.
3. Unavailability of relevant information on program performance.
4. Inability of policymakers or managers to act on the basis of evaluation information (p. 37).

Wholey et al. goes on to outline key steps to be taken to resolve these issues. Rossi et al., (1999) describe the activities of evaluability assessment as: "(a) description of the 'program model' with particular attention to defining goals and objectives, (b) assessment of how well defined and evaluable that program model is, and (c) identification of stakeholder interest in evaluation and the likely use of the findings" (p. 157). From these two descriptions, it is seen that evaluability assessment is a large part of the first intervention research task (Gather Background Information/Conduct Needs Assessment) in the developmental research phase of IER. One important advantage of evaluability assessment is that it provides a structured process for describing and assessing a program's conceptualization of what it is supposed to be doing and why (Rossi et al., 1999). Goldenhar et al. (2001) have commented that sound developmental and implementation studies could provide much of the needed data for determining if and when a given program is worthy of a formal and potentially expensive effectiveness study.

Traditional Farm Safety Education

One final point about evaluation should be addressed. As the information in this and the preceding sections on program evaluation makes clear, a high quality intervention evaluation requires considerable expertise, resources, and effort. Agricultural safety and health professionals may be involved in literally dozens of activities, projects and programs that are intended to help reduce farm hazards and risks, and to increase safety knowledge, attitudes and practice. Setting aside for the moment formal evaluability issues, it is a practical reality that not all of these efforts are worthy of formal evaluation effort. For example, countless slide sets, filmstrips, videotapes, fact sheets, brochures, pamphlets, publications, exhibits, and demonstrations, often presented irregularly and with other unrelated material, marks the traditional generalized approach to farm safety education (Murphy, 1992). A question that can be asked is, "Should such educational endeavors be discontinued?"

This question is asked from the perspectives of farm safety cultural values, and public support for agricultural safety and health. In lay language, culture is the way of life of a community or group (Taylor, 1976). A value is something that is desired or prized. It follows that cultural values refer to those things that persons who share a common culture hold in high esteem. In turn, cultural safety and health values refer to the ways those persons who share a culture think and act about safety and health (Murphy, 1992). Increasing farm work safety and health as a cultural value would likely be inhibited by any widespread action that reduces its visibility. That is, it would seem difficult to increase the cultural value of farm work safety and health among the farm population by exposing them to less of its recommended practices. In a similar vein, maintaining or increasing local, state or national support for farm safety and health may become more difficult if it becomes less visible at each level because good evaluation isn't a part of a given effort on behalf of farm safety and health.

Support for farm safety and health as a prized cultural value among the farm population, in part, is gained through the collective and public promotion of farm safety and health as it is done through the thousands of small, dispersed, informal and one-shot presentations across the country. A goal of theory-driven evaluation, which is encouraged by the IER, is to efficiently move to more effective interactions with the targets of injury prevention programs, be they farm audiences, agricultural educators, community leaders, academic researchers, political supporters, or funding sources. The problem to be avoided is letting farm safety and health lose its already limited cultural value as attempts are made to move toward more effective intervention practices.

Surveillance of Agricultural Risk and Injury Incidents

The third theme area revolves around agricultural risk and injury surveillance. Surveillance refers to the ongoing systematic collection, analysis and reporting of data for use in planning, implementing and evaluating practices and programs of public health importance (Thacker, 2000), and is considered an essential element of public health practice (Halperin, 1992). The purpose for surveillance is sometimes confused with the purpose for etiologic or epidemiological research. Primary roles of surveillance include identification of problems, describing the scope and extent of problems, and evaluation of interventions designed to ameliorate the problems. This data is most often presented as a rate of injury over a specified time frame, and as descriptive categorical data. The primary roles of epidemiological research are to determine necessary and sufficient causes of problems, in this case work-related injury identified through surveillance data, and to guide intervention development (Timmreck, 1994). Therefore, epidemiological research activity should follow surveillance activity. Nevertheless, surveillance and epidemiological research roles and activities often become blended. One reason for this is that the same descriptive data collected for surveillance purposes can also be used for epidemiological research. The more detailed or the wider the variety of descriptive data collected, the more likely it is to serve the purposes of both surveillance and epidemiological research.

Another reason that this may occur is because surveillance data is often referred to as a "cause" of a particular type of incident. For example, the source or thing that is identified by surveillance as inflicting an injury is often referred to in casual conversation as the "cause of the injury." In agriculture, general surveillance data that may be identified as a cause of injury includes machines, animals, work surfaces, structures, chemicals, dust, trees, heat, and the like. From an epidemiological research perspective, the cause of any injury incident is usually much more complicated than is conveyed by reference to a single item or agent. The cause of agricultural injury incidents, from an epidemiological research perspective, is beyond the scope of this section, although it is obviously connected. It is discussed in this section only in the context of its use with surveillance data.

Surveillance for agricultural risk and injury has been occurring for decades, but a major challenge for the future is to improve its quality, consistency, and usefulness. General risk exposure and descriptive injury data abounds in the literature, but its utility in moving an understanding of agricultural work risk and injury beyond that of 20 years ago can be questioned. Arguably, the current national picture of agricultural injury, with perhaps the exception of tractors and ROPS issues, is ambiguous and contradictory. For example, a recent comprehensive review of agricultural injury research by McCurdy and Carroll (2000) showed: (a) a very large range between low and high overall injury risk (.5/100 person years to 16.6/100 person-years, the highest rate being more than 33 times the lowest rate); (b) that there is increased risk among younger persons (<30 years of age), persons born after 1940 (approximately <54 years of age) and older persons (>55 years of age); and (c) that nearly every group of agricultural workers is at "increased risk of injury." Part-time hired workers, full-time hired workers, farm family workers, farm resident workers, nonfarm resident workers, farm owners, farmers with small farms, and farmers with large farms, are all groups that researchers have shown to be at increased risk of agricultural work injury.

For practical reasons, stemming primarily from limited personnel and financial research resources, and the extensive variability of agricultural work, workplaces, and workforces, it may be that risk and injury data collection for agriculture will always need to serve the dual roles of surveillance and epidemiological research. Given the need of the surveillance purpose to precede the epidemiology research purpose—accurately identify and describe problems before attempting to establish their causes—the quality of the data collected, defined here as consistent, insightful, and informative, should be of prime importance. However, an argument can be raised that searching for more precise causes of agricultural injury, through the collection of more detailed or wide-ranging data is not the most efficient use of limited resources. This argument will be returned to later in this section.

Regardless of the perspective on the depth and range of the risk and injury detail needed, injury intervention effectiveness begins with clear identification and description of the problem. The discussion now turns to this issue. Most agricultural safety and health professionals would agree about the need for clear problem identification and description, as expressed by risk exposure and descriptive injury incident data. As noted above, and as will be further explained, the picture of agricultural injury is cloudy at best, and the United States does not appear to have a strategy for addressing this important and basic issue. The lack of clarity of agricultural injury data, and the absence of an ongoing national plan to improve agricultural risk and injury surveillance, are embedded in the factors and characteristics as described in "The Nature of Farm Work" section above, and are not easily resolved. But these factors must be overcome if we are to see more clearly how we might significantly decrease the rate of farm work related injury in the next decade. Contemporary issues regarding agricultural injury data are presented below. This discussion includes a short history of agricultural injury surveillance as a perspective for current problems and issues.

Risk Exposure and Injury Rates

Farm work injury rates based on numbers of farms, farm workers, and farm residents have long served as surrogates for actual exposure to farm work hazards. Actual exposure to specific hazards is conceptually the preferable measure of exposure (Ruser, 1998), but as agricultural safety and health researchers know, this can be very difficult to obtain in agriculture. Many national studies over the years have documented broad levels of farm work injury risk based on an annual number of accidents/injuries per numbers of farms, farm workers, or farm residents. For instance, the Bureau of Agricultural Economics (1948) reported on injury rates per 100 farms. Rush (1957), Fritsch (1976), and Fritsch and Zimmer (1980) have all reported fatality rates per 100,000 farm residents, while Fritsch and Zimmer (1980) and Myers and Hard (1995) have reported fatality rates per 100,000 workers. Runyan (1998) reported fatal and nonfatal injuries per 1,000 farms, and Hard et al. (1999) and Myers et al. (1999) published fatality rates per 100,000 workers less than 25 years of age, and more than 55 years of age, respectively. These rates are estimates in addition to the annual fatality rate estimates per 100,000 workers published annually by the NSC and BLS, and periodically by NIOSH (see Figure 1.1 in Chapter 1).

At the state level, the Pennsylvania Department of Public Instruction (1957) has reported injury/accident rates per 1,000 farm residents and per 100 farms. Wardle (1958) reported the number of fatal accidents to Iowa farm people per 100,000 farm population. Phillips et al., (1975) reported accidents per 1,000 Ohio farm residents. State reports that used the Standardized Farm Accident Survey program (discussed in more detail below) reported on accidents /injuries per 1,000 population (Paterson et al., 1972) and per 100 farms (Murphy, 1977). More recent reports include those by Pratt et al. (1992) reporting the number of injuries per 1,000 central New York dairy workers, Browning et al. (1998) reporting on injuries per 100 farmers to older Kentucky farmers, and Liller et al. (2000) reporting on Florida adult fatal injury deaths per 10,000 farms. A regional report on tractor-related injuries in five states (Minnesota, Wisconsin, South Dakota, North Dakota, Nebraska) reported tractor injuries per 100,000 farm household population and per 1,000 farms (Lee et al., 1996). This report contains numerous rates of injury by categorical variables that are traditional (e.g. size of farm, age of operator, hours of work) and novel (e.g. marital status, relationship within household).

Broad rates such as work injury incidents per number of farms, farm workers or farm residents have value as a statement of work risk at a single point in time, and over time indicate whether risk is decreasing, staying the same, or increasing. Assuming that prevention of injury to a person in the agricultural work environment is the main goal, then it follows that a person-based rate has more value than a farm-based rate. There are two common ways of expressing agricultural work person-based rates: (1) number of workers, and (2) number of residents. From an occupational work perspective, a worker-based rate has more value than a resident-based rate because not all farm residents participate in agricultural work. Including non-working farm residents in a work-based rate would result in an understatement of risk because more people would be counted in the denominator value.

Even a person-based rate has its limitations. Conceptually, it is not difficult to understand that the most precise measure of person-based risk is based on actual exposure of an individual to a particular hazard or activity, and that number of hours is a convenient and standard unit of measure. On the other hand, the sizable variability in the types and sizes of agricultural operations, of agricultural work, and among the people doing the work, makes collecting individual hourly exposure data onerous, and this is why surrogate measures are used. Ruser (1998) has investigated the effect of using the number of workers (employment) as a proxy for worker exposure to risk in the calculation of fatality rates. Using data from the Census of Fatal Occupational Injuries (CFOI) and annual employment and annual hours worked data from the Current Population Survey (CPS), he developed 100,000 worker-years fatality rates with employment and hours worked as denominators. In general, he found that the rank correlation of employment- and hours-worked-based fatality rates for 222 occupations was highly correlated (.99). That is, each method revealed a very similar degree of risk for the same occupations and for most groups of workers. However, some important differences were found when he calculated fatality rates by age groups. Specifically, he found significant differences between the two rates with the youngest age group (15-19 year olds) and the older age group (ages 65 and over) of workers. This finding has obvious significance to agriculture.

Ruser (1998) calculated fatality rates per 100,000 worker-years by gender, age, occupation and industry. For all groups except young workers and older workers, the difference between the two methods of calculation was 10% or less. For example, the fatality rate for men using employment as the denominator was 8.7 per 100,000 worker-years, while it was 8.0 using hours worked as the denominator. For the youth group, however, the difference between methods was 60% (2.5 vs. 4.0), while the difference was 37% for older workers (14.0 vs. 19.2). In both cases, using employment as the denominator indicated less risk for the group than using hours worked as the denominator. With these exceptions, the number of workers appears to be a reasonable surrogate for hours of work in the calculation of work fatality rates.

Nevertheless, a rate based on a total employment number cannot communicate risk or other information as accurately as a rate based on exposure hours because the relative value of an employment number across a group differs from the relative value of total hours worked for the same group. For example, workers who work part time will show as a greater fraction of total employment than of total hours worked (Ruser, 1998). To illustrate, if five workers of a 10-person workforce are working part-time, e.g., 20 hours per week rather than 40 hours per week, those part-time workers account for 50% of the total workforce, while accounting for 33% of total hours worked per week (100 hrs. of 300 hrs.).

Surveillance Efforts

Agriculture has a long history of efforts to collect work exposure-based injury data. The first published report found of this being done was when the Pennsylvania Department of Public Instruction (1957) reported a farm work accident rate per 1,000,000 hours of farm work (including fatal and nonfatal cases). The department then compared the injury rate per million hours worked in farming with other major industrial groups and found farming to be slightly more hazardous than any other industrial group, including lumbering, coal mining and construction. The report also noted accident rates per million hours worked by size of farm operation and found smaller farms (1-199 acres) had a much higher accident rate than larger farms.

Statewide surveys took a major step forward in sophistication in 1967-1968 when Ohio and Michigan used similar methods to stratify and randomly select farms for inclusion in the survey. Similar data collection forms and procedures were also used, and hours of exposure (hours worked) data were collected. The Ohio procedures and data were described in articles by Stuckey and Pugh (1973) and Philips, Stuckey and Pugh (1975). Hofmeister and Pfister (1968) and Pfister and Hofmeister (1969) published the Michigan procedures and data. Based largely on the experiences from the Ohio and Michigan surveys, the NSC published a handbook outlining standardized definitions, procedures and survey forms for use by states in collecting a statistically valid statewide sample of farm injury data (Hanford and Conrath, 1972). A national strategy for states to implement the Standardized Farm Accident Survey program was devised, with state leadership provided primarily by cooperative extension and Farm Bureau. The goal of the survey program was to develop a valid nationwide picture of farm work injury and illness. This goal was not entirely reached as some types of farms were overrepresented (e.g., farms more than 200 acres, dairy farms) and some were underrepresented (e.g., farms less than 100 acres, beef farms) (Hanford et al., 1982; Hoskin et al., 1988).

NSC began publishing pooled data from standardized surveys in 1972 (Hanford & Conrath) with an eight-state report, and concluded with a 1988 contracted report to NIOSH that featured data from 35 surveys in 31 states (four states repeated the surveys) (Hoskin et al.). The 1982 report that covered the 31 states (no repeat surveys) was the last, most comprehensive report that was widely distributed (Hanford et al.). Occupational illness data was collected along with the injury data, but there were an insufficient number of illness cases to make national estimates. The 1982 *Farm Accident Survey Report* included the injury data from 31,398 farms and provided a national and regional epidemiological picture of farm injury. Work injury rates per million hours of exposure by age group, sex, worker status, agricultural operation and size of operation were published along with numerous tables, including cross-tabulations, showing how the percent of injuries were distributed among many variables (Hanford et al., 1982; Hoskin et al., 1988). Additionally, several extension specialists produced state-level reports containing farm work injury rates based on hours of exposure. A few examples include Stewart (1977), Brazelton et al. (1976), and Williams (1983).

Among the important findings from the standardized 1982 *Farm Accident Survey Report* were:

- A national accident rate of 20.5 per million hours worked, with regional rates (approximately 5 contiguous states per region) that varied between 9.4 and 24.6;
- Per million hours worked, hired labor had a higher rate of injury (22.5) than family workers (20.7), beef farms had a higher rate of injury (20.0) than fruit (17.6), grain (17.6) or dairy (17.4) farms, and 5- to 14-year-old workers had the highest rate of injury (24.0), 15 to 24 year olds the second highest (23.4) and workers 65 and over had the lowest rate (17.3);
- More than 40% of the victims had less than one day's experience with the "thing involved" in the injury while 20% had more than six months experience;
- More than 65% of the injuries were considered "severe" (e.g., broken bone, sprained back) or worse;
- Animals, machinery and trucks/other vehicles were the leading items involved in farm injury incidents (16.8%, 16.3%, 15.3%, respectively), but machinery incidents resulted in more permanent injuries (51.5%) than all other sources combined;
- The tractor, while involved in only 8% of all incidents, was involved in 33.3% of fatal incidents, with truck/other vehicle and machinery following at 19.1% and 19.0%, respectively; and
- Three specific activities at the time of injury, chores (25.4%), field work (21.1%) and machinery maintenance (12.7%), accounted for nearly 60% of the injuries.

A study in 1971 also contributed valuable actual exposure-based farm injury data. Doss and Pfister (1972) published a study that has provided a detailed understanding of farm work injury risk. This study, based on actual hours of use of tractors and machines, showed:

- Tricycle tractors were approximately 52% more likely to be involved in accidents than wide front-end tractors;
- Tractor operators in Michigan 10 to 14 years of age and 65 years and older were nearly 9.5 and 6.5 times more likely, respectively, to be injured while operating a tractor than operators 25 to 44 years of age;
- 5- to 14-year-old tractor operators in Michigan and Ohio were nearly 5 and 3.5 times, respectively, more likely to be involved in a fatal tractor accident than 25- to 44-year-old tractor operators; and
- That per hour of use, elevators (for moving produce) were the most hazardous machines, followed by combines, wagons, corn pickers, and all other machines except the tractor.

Other major reports and publications from the 1970s added insight to the national and state pictures of farm injury and injury risk. For example, a 590-page U.S. Secretary of Transportation report to Congress in 1971 helped to amplify hazards and injuries associated with tractor operation. And the U.S. Bureau of Labor Statistics (BLS) published agricultural injury and illness rates in its annual report of occupational injury and illness prior to 1976 when Congress passed the record-keeping exemption for small employers, and continued to publish rates for farms with 11 or more workers after 1976. So, by the early 1980s, considerable data and other information on injuries and hazards existed that described the who, what, when, where and how of farm work injury.

In 1997 and 1998, Myers updated exposure-based farm work injury data by publishing several injury rates based on national surveys designed by NIOSH and conducted by USDA in 1993 and 1994 through an interagency agreement. The 1994 data, published in 1998 (Myers), is cited here. The surveys established national, regional and state estimates of lost-time injury rates and restricted workday rates; and on national and regional levels, estimates of injuries by several categorical variables, such as worker status, gender, age group, etc. In computing national rates of lost-time injury by farm type, Myers found nurseries to have the highest lost-time injury rate per 200,000 hours of work (9.4); followed by dairy operations (6.8); vegetable, fruit and nut farms (5.0); and beef, hog and sheep operations (4.6). But when examined by restricted workdays—a surrogate for severity of injury—the order is much different. Field crop operations had the highest rate of restricted workdays per 200,000 hours worked (122.5); followed by beef, hog and sheep farms (119.0); cash grain operations (111.9); and vegetable, fruit and nut farms (99.0). Myers was also able to produce national and regional rates of injury by type of farm operation and worker status. He found, for example, that hired workers on nurseries had a much higher rate of injury (11.8) than family workers (4.2), and that family workers of dairy operations had the highest injury rate of family workers for all types of operations.

Castillo et al. (1999), using procedures similar to those identified in the discussion of Ruser's work (1998) and earlier work (Castillo et al., 1994), updated the statistics on youth work-related fatal injury in agriculture. Several fatality rates per 100,000 full-time equivalents were presented comparing youth workers with adult workers. A full-time equivalent (FTE) is computed by multiplying the average number of hours worked per week times 52 weeks and dividing by 2,000 (the average number of hours worked per calendar year). The FTE is then multiplied by 100,000. Among the findings were:

- A five-year average of 12.2 deaths per 100,000 FTE;
- Rates per 100,000 FTE for youth are comparable to rates for adults up to the age group of 45 to 54 years;
- Crop production had the highest fatality rate (20.8) among 15 to 19 year olds; and
- Rates were similar across regions (Midwest, South, Northeast, West) but were different by agricultural sector within regions. For example, the Midwest region had the highest rate (35.0) of fatal injury in the crop production sector while the Northeast had the lowest (9.6). And the Northeast and West regions had the highest rates (both 13.2) in the livestock sector while the Midwest and South had equally low rates (6.7).

Descriptive data on occupational health hazards and associated illness or disease is not as easily documented as is the data on traumatic injury, and few studies with rates of occupational agriculture disease or illness have been published. While basic occupational health hazards of farming (such as hazards associated with pesticides, fertilizers, sun and heat, zoonoses, pulmonary diseases and mental health) have been identified since at least the 1950s (Axelrod, 1949; Top, 1956; Berry, 1959), new insights into the extent and seriousness of such hazards were still being documented in the 1970s and 1980s. Examples of this are articles on the effects of pesticides (Yoder et al., 1973; Wickes et al., 1979), prevalence of Farmer's Lung (Madsen et al., 1976) and hazards from swine confinement operations (Donham & Gustafson, 1982). Nevertheless, by the end of 1980s considerable descriptive epidemiological information did exist. The papers from the second international symposium in 1985 in Saskatoon, Canada (Dosman & Crockcroft, 1989) and an article by Cordes and Rea (1988) provided good reviews of occupational health hazards in agriculture in the mid-1980s.

In the 1990s, farmer reports of high levels of job stress and evidence that portions of the agricultural workforce may be suffering higher than normal rates of stress-related diseases gained attention. Preliminary data on perceived stress, cardiovascular diseases, stroke and other health indicators all suggested that rates were elevated among managers of small farm operations (Chapman & Scharf, 1994). Some research on the incidence of suicide, particularly in relation to economic downturns in the agricultural industry, also appeared to provide a link between occupational stress and farming (Gunderson et al., 1993).

Cumulative trauma disorders (also known as musculoskeletal disorders) are another hazard of the agricultural workforce that received attention in the 1990s. Generally not considered traumatic injuries, musculoskeletal disorders include problems like chronic pain in the wrist, knee, hip or back. A number of production agriculture occupations, especially those with repetitive or forceful physical activities (e.g., milking cows, manual field labor), appear to place individuals at risk. Work is ongoing to determine the extent and severity of this risk throughout production agriculture (e.g., Baron et al., 2001; Estill & Tanaka. 1998) and to develop and evaluate interventions (e.g., Baron et al., 2001). Overall, descriptive data on occupational agriculture health hazards and injuries expanded and became more detailed during the 1990s through the publication of major references such as the 617-page book, *Agricultural Health and Safety: Workplace, Environment, Sustainability* (eds. McDuffie et al., 1995); the 758-page book, *Safety and Health in Agriculture, Forestry, and Fisheries* (eds. Langley et al., 1997); and the 416-page compilation of selected papers from the Fourth International Symposium "Rural Health and Safety in a Changing World," Saskatoon, Canada, by the *Journal of Agricultural Safety and Health* (eds. Murphy et al., 2000).

Based upon the literature cited in this section and the preceding two chapters, it is evident that broadly representative state and national occupational agriculture risk exposure and injury incidence and prevalence data has existed since at least the early 1970s. Therefore it is fair to say that agricultural safety and health professionals have had an abundance of surveillance data and information that have generally identified a great deal of the who, what, when, where and how of farm work-related injury and illness for nearly 30 years.

Identifying Agricultural Work Injury Incidents

The discussion so far has revolved around the denominator or bottom portion of agricultural work injury rates, primarily the merits of different denominators used to express work-related injury risk and the history of collecting and publishing these rates. This bottom or denominator portion reflects the size of the workforce or hours of work in the injury rate calculation. The numerator, or the top portion of the rate figure, is also problematic and has been from the beginning of agricultural injury surveillance. Problems arise from a variety of definitional and terminology issues. These issues have been previously discussed in the literature and will only briefly be reviewed here (e.g., see Purschwitz, 1992). Most agricultural researchers understand these issues and many attempt to account for them in their research. But the farm population, from which most hazard exposure and injury data is obtained through self-reports, has less appreciation of the subtleties of research definitions and terms, and this in turn may mean less reliable identification of the cases of interest on surveys. Local, state and national officials and researchers responsible for injury reporting are another group that often misclassifies agricultural injuries due to their lack of appreciation of definitional and terminology issues.

For example, farm "workplaces" are not easily defined because the business of farming can take place anywhere, including in the home, at a restaurant, or at a stop along the way while taking children to school. A hazard exposure or injury incident may not be remembered when farm work occurs away from more traditional farm work locations. Farm research participants may interpret the term "agricultural injury" in different ways. They may think only of incidents that resulted in immediate, serious injury, and not count an incident for which medical treatment was not immediately sought. Or they may include any injury incident that happens on the farm, including recreational, lifestyle, and nonoccupational incidents for which the researcher has not asked.

Researchers and research subjects alike can have different interpretations of the phrase "farm work-related injury." For example, one interpretation is of injury only while the person is actually working. This would exclude bystanders or visitors injured by something in the work environment but who were not working themselves. A second interpretation may be anyone injured during a work activity. This would include the victim or visitor injured by exposure to the work environment even if he or she was not working, but it excludes victims injured in a work environment when no work is being performed. Therefore, a third interpretation may be that anyone injured or any incident that occurs from exposure to a farm work environment should be included in the data. But this may still miss some who are injured when the victim's primary occupation is not farming. Finally, some consider intentional injury, such as a suicide or an assault, as a work-related injury if it occurred in the course of work or as a result of it, while others do not.

Case identification can vary dramatically from different interpretations of the same material. For example, Conroy and Sciortino (1997) found that, of 504 cases in the California Occupational Mortality System (COMS), either 258 cases, 306 cases, or 104 cases of fatal agricultural work-related injury were identified, depending on the specific case definition used to search the database. Murphy et al. (1990) found, from a possible 66 cases in Pennsylvania that were farm related, either 26 cases or 39 cases that could be ascribed to occupational agriculture, depending on the case definition used. In a second project, Murphy et al. (1993) reclassified 101 Pennsylvania cases that were originally classified as occupational agriculture using the proposed Farm and Agricultural Injury Classification (FAIC) code, and found that only 60% of the 101 cases still remained within the occupational agriculture class. In a third project in Pennsylvania, Yoder and Murphy (2000) found they were able to more accurately judge the actual occupational work status of victims in 51 of 90 cases that had been previously coded using the FAIC code. The underlying issues in all of these studies were how "occupational work" is defined or interpreted, and how much information is available about the case. Establishing a consistently interpreted and used definition of occupational work within the industry of agriculture, forest and fishing remains problematic due to: (1) the overlap of farm residences and workplaces, (2) more than one occupation by many who operate farms or perform farm work, (3) the use of family and voluntary labor, (4) the use of seasonal and migrant laborers, and (5) a reliance on the farm population to often supply case information. In addition, two more reasons for misclassification and confusion can be identified.

First, while acknowledging the CFOI system improves and standardizes case definition for groups that officially report on occupational injury (e.g., the Bureau of Labor Statistics, NIOSH, and the NSC), by definition the CFOI system only includes fatal cases. There is no equivalent inclusive surveillance system for nonfatal cases in agriculture. Second, changes between the Standard Industrial Classification (SIC) code and the new North American Industrial Classification System (NAICS), which assigns business establishments to broad industrial categories, changes the makeup of the agricultural, forestry and fishing industry. The change most relevant to this discussion is that commercial logging operations are now considered a part of the agricultural, forestry and fishing industry. They were previously in the manufacturing sector. Wood lots and stands of trees are common on farms in many states (e.g., throughout the Northeast and several states in the upper Midwest), and whether an injury incident involving the cutting and clearing of trees and woods was an occupational or nonoccupational activity has always been difficult for reasons discussed earlier. Throwing commercial logging into the mix does nothing to help clarify surveillance findings in general or to standardize case definition issues. While it will be possible to accurately separate commercial logging cases from production agricultural (farming) cases in official reports (e.g., BLS, NIOSH), this is not as likely among farmers, farm workers, or university researchers. This lack of accurate separation of commercial logging cases from farming cases by farmers, researchers, or anyone for that matter, will only add to reliability and validity problems with case identification and analyses in occupational agriculture injury reports.

Categorical Data Inconsistencies

If hazard exposure and descriptive injury data should guide injury prevention programs (Thacker, 2000; NIOSH, 2001b), then the inconsistency with which agricultural researchers collect hazard exposure and descriptive categorical data is particularly troublesome because the published literature is beset with categorical variables that are defined differently. Because precise exposure data is difficult to collect, having some consistency—and therefore comparability—among different studies becomes all the more important as an aid to identifying risks and hazards and subsequent injury prevention actions. Most agricultural safety and health researchers and educators are familiar with this lack of consistency in surveillance data so only a few examples from the literature will be provided.

Age groups of injured victims are often presented and described differently. For example, some researchers present data using five-year age groups (Gerberich et al., 1993), 10-year age groups (Napier et al. 1985), 20-year age groups (Stueland et al., 1990), or mixed age groups, such as 22 years, 16 years and 29 years (Wilkins et al., 1999), and some only provide the mean age of injured victims (Mills et al., 1989). Osorio et al. (1998) used mixed age groups for lower (<18 years) and upper age groups (ages 61-64, 65+) and 10-year age groups for the middle ages, but used less traditional age groups of 21 to 30, 31 to 40, etc.

Researchers use different ages to describe minors (youth, children, adolescents) and older (aged, seniors) injured or study subjects. For example, the young group is sometimes defined by the ages of 1-4, 5-9, 10-14 and 15-17 (Liller, 2000); 0-5, 6-15 and 16-18 (Stueland, et al., 1991); =/<13, 14-15, 16-17 (Heyer et al., 1992); <10 and 10-19 (Lee et al., 1996); less than 20 years of age (Rivara, 1997); and less than 25 years of age (Hard et al., 1999). Similarly, the older group has been defined as age 55 and older as a single group (Myers et al., 1999); as 55 and older by five-year age groups with age 70 or more the oldest category (Browning et al., 1998); 60 and older as a single group (Gelberg et al., 1999); 65 and older as a single group (Mills et al., 1989); not specifically defined but reasonably inferred as age 60 and above (Lee et al., 1996,); or left undefined (Crawford et al., 1998).

Another categorical variable whose definition seems nearly arbitrary in the agricultural safety and health literature (contributing to confusion and lack of comparability among studies) is the category that attempts to identify the main object or substance involved in an injury incident. This variable has been termed "agency of accident" (Hoskin et al., 1988), "main object involved" (Mills et al., 1989), "types of farm practice involved" (Cummings, 1991), "device" (Stueland et al., 1991), "source of the injury" (Myers, 1998), and "external cause of injury" (Zhou & Roseman, 1994) among other things (e.g., older reports used variants of "type of," "kind of," or "thing involved"). Although experienced professionals may be able to sufficiently interpret these various labels for research and education purposes, there is no way to interpret—and therefore compare—studies when the variables that make up this category (no matter how it is labeled) are different. For example, Napier et al. (1985) listed 13 causes of injury, Hoskin et al. (1988) listed nine, Stueland et al. (1991) listed nine, Murphy and Yoder (1998) listed 11, Myers (1998) listed 12, and Crawford et al. (1998) listed 13. Even when the same number of causes/sources appears, the list can be different. For example only six of the categories overlap in the Hoskin et al. and Stueland et al. lists, and in the lists of Napier et al. and Crawford et al., only seven variables are the same.

Even a single state does not always report identical variables for the same data set. For example, the Injury Epidemiology Division of the Oklahoma State Department of Health (1991) reported cause of death for 1980-1989 by three different sources and used five different variables, while reprinting state medical examiner records for the same time period that showed 15 different cause of death variables—and then only two variables (drowning, farm machinery/machines) were identically defined.

Some researchers have separated "type of event" from "cause/source of injury" but others have not. For example, Osorio et al. (1998) and the NSC (2000a) have some type of event variables (e.g., struck by, struck against) within the cause/source of injury category, while Hoskin et al. (1988), Myers (1998) and Lewis et al. (1998) have separate categories. Additionally, Hoskin et al., Myers, and Lewis et al. each used six variables to describe "activity at time of injury," but only two variables (field work, treating/handling livestock/animals) are common to all three studies. Even farm type and a single-injury source, e.g. a tractor, have been described inconsistently. For example, Hoskin et al. (1988) listed five types of farm operations, Myers (1998) listed eight and Browning et al. (1998) listed four. With tractor injury incidents, Myers et al. (1998) described eight types of tractor-related fatality events while the NSC (2000a) reported four.

Similar disparities can be found among almost all the categorical variables used to describe farm injury and hazard exposure information. The issue here is not which makeup of categorical variables is correct, or even the best. Rather, the issue is consistency for comparability of studies so that surveillance data is consistent and so that researchers and educators may communicate with each other effectively, and also with all external audiences (e.g., farmers, county agents, vocational agriculture teachers, the media, and policy makers).

Nonfatal Injury Surveillance

Closely related to inconsistent data categories is the issue of periodic or absent surveillance of nonfatal agricultural injury. Whether at a state, regional or national level, the United States does not have a plan for regular, periodic surveillance of nonfatal agricultural injury. Recently, four plans have been implemented that have provided some level of a state, regional or national picture of nonfatal agricultural injury: the NSC's Standardized Farm Accident Survey Program (Hanford & Conrath, 1972); The BLS's Survey of Occupational Injuries and Illnesses (US BLS, 2000); the U.S. Census of Agriculture survey in 1992 and 1997 (USDA, 1992, 1997); and NIOSH's Traumatic Injury Surveillance of Farmers Survey (Myers, 1998).

The NSC no longer reports data on nonfatal farm-related injury. Reasons for the demise of the NSC's survey program included finding a representative sample of participant farms and a heavy reliance on volunteers for data collection (Hoskin et al., 1988; Murphy & Huizinga, 1989). The BLS survey has excluded the majority of farm operations since 1976 because of congressional restrictions. The Census of Agriculture survey collects only limited data (two questions) on a noninclusive population of workers (excluding contract workers and custom workers), so it is of limited usefulness. It is not assured that even this limited surveillance will be included on the 2002 Census of Agriculture survey. NIOSH's survey, originally conceived of as a periodic, national surveillance program, was conducted from 1993-1995. It has not been conducted since then, although there are agency plans to reconduct this survey (Myers, 2001). Clearly, nonfatal injury surveillance that is systematic, ongoing, representative, and inclusive has been and remains an important surveillance problem.

The Standardized Farm Accident Survey program was an ambitious surveillance project. It was designed to collect numerous demographic details on every member of the farm family and about the farm operation, and the total hours worked by family and hired workers. A general accident and illness report form was to be completed for each incident. There were then 13 supplemental data forms for obtaining additional details as appropriate. For example, a medical and cost information supplemental form asked about workdays lost and medical costs; a tractor supplemental form asked about the type, size, age, etc., of tractors; and another supplemental form asked for details about the animals and environmental conditions present at the time of injury, etc. However, the survey program's reliance on volunteer data collectors ultimately proved to be a major factor in why it did not continue (Murphy & Huizinga, 1989).

The other three surveillance programs are government-based efforts where volunteer data collectors are not an issue, however, political considerations are. For example, the exclusion of family farms from the Occupational Safety and Health Act of 1970, and the restriction of enforcement of its regulations on small farm operations since 1976, was a political decision. These decisions, no matter how well or illfounded, have prevented and impeded nonfatal injury surveillance in agriculture. The important point about any government-based injury surveillance program or system is that it is inherently tied to political issues and decision-making that may be far removed from the intrinsic value or need for the surveillance activity.

Data Detail and Depth

The abundance of general agricultural risk and injury data has been documented throughout this and the previous chapters. It is common, though, for researchers to write about the lack of, or inadequacy of, agricultural risk and injury data. For example, McCurdy and Carroll (2000) wrote, "Epidemiological investigation of health status among agricultural populations is frustrated by inadequate data for number of persons at risk and the number of persons suffering injury" (p. 464). Presumably, McCurdy and Carroll and others that have expressed similar sentiments throughout the 1990s were writing about the lack of specific, in-depth risk and injury data. Earlier a question was raised about the need for more detailed or in-depth data. For example, would data that identifies the rate of work-related injury by a combined injury source by major work activity substantially increase an understanding of injury incidents and/or ability to develop effective interventions? At least three broad arguments can be raised for questioning the use of limited resources to obtain substantially more detailed and in-depth agricultural risk and injury data through surveillance activity. These broad arguments cannot be definitively argued here one way or the other, since they are complex and intertwined with many other issues. Nonetheless, the purpose in raising these broad arguments is to help stimulate an open discussion on them as they relate to priorities for future agricultural risk and injury surveillance efforts.

The first argument is practicality. Agricultural work is almost infinitely variable. Therefore, the possible combinations of factors contributing toward any single exposure or incident are also infinitively variable. To gain any significantly new insight into the circumstances surrounding risk exposures and injury incidents, broad categorical variables—and combinations of variables—would have to become so specific that survey forms would likely become very long and probably unwieldy. This leads to nonresponsiveness by survey subjects to questionnaires. It is telling that the designation of "other" is often a leading activity and source of injury variable in general agricultural injury reports (e.g., Hanford et al., 1982; Myers, 1998; Purschwitz, 1997) and even in more specific reports. For example, in a report by Lee et al. (1996) on regional tractor-related injury incidents, "other" was the second highest reported activity at the time of injury event, after mounting and dismounting. The "other" category is used in injury reports as a matter of convenience to combine activities or injury sources when each activity or source individually contributes a small portion to the total. This supports the notion of extensive variability in agricultural work activity and sources of injury. It may be that the situational exposures that come together and result in an agricultural injury are so varied, they are meaningful to almost all observers only when grouped in general categories.

A second reason comes from a sociopolitical perspective. In 1991 at the Surgeon General's Conference on Agricultural Safety and Health, James Merchant spoke about the lack of safety and health regulation in agriculture and the need for innovative intervention models. Using, as an example, the need to get the proven intervention of ROPS on all new and used tractors, he stated, "If we cannot develop a U.S. model for a proven intervention on the single most important cause of agricultural mortality, how can we succeed in addressing less dramatic yet still important causes of agricultural diseases and injuries?" (Myers et al., 1992, p. 74). This question sums up a great deal about the state of agricultural injury prevention in the United States. As a society we are unwilling to impose by law on all farmers an effective intervention—a ROPS on every tractor—for the most serious work-related injury problem in agriculture. We also know that keeping children, particularly those less than 14 years old, out of the farm work environment would save the lives of scores of children each year. But we choose as a society to not impose by law such an intervention on farm families or farm operations. The issue for now isn't why we make these choices, but rather what value is there in collecting more detailed or in-depth data on problems, for example on tractor rollover incidents or on children being in a farm work environment, when the real intervention barrier is social and political acceptance, rather than lack of data.

A third broad argument comes from injury epidemiology. Two widely-acknowledged principles of injury epidemiology are that: (a) the selection of a preventive or control strategy is not dependent upon the rank order or importance of casual factors, and (b) the prevention of injury does not depend upon an exhaustive understanding of injury causation. For example, ROPS protect tractor operators from a crushing injury regardless of why tractors rollover. A complete enclosure (guarding) of a power take-off (PTO) driveline shaft will protect a person standing next to an operating driveline—regardless of why a person would stand next to an operating driveline. A tractor running at night on a busy highway with no visible SMV emblem or rear flashing lights is a serious hazard and is prohibited by law, without any need for understanding exactly why a farmer would take such a chance in the first place. There are many hazards in agriculture that can be corrected through design innovations, environmental and worksite controls, or regulations, without exhaustively knowing or understanding the details that contribute to the occurrence of an exposure or injury incident.

Improving Agricultural Injury Surveillance and Research

There is no simple solution to improving agricultural injury surveillance and research. Nor are agriculture injury professionals alone in struggling with how to better plan, coordinate and implement injury surveillance systems. For example, NIOSH (2001b) has published a strategic plan for hazard and injury surveillance to help guide its surveillance efforts over the next several years. Enhancing surveillance of occupational illnesses, injuries, and hazards in agriculture is listed as one objective. More than 400 health and safety professionals contributed to the development of the strategic plan.

A few ideas are offered in Table 4.2 as a starting point for beginning a dialog on improving agricultural injury surveillance and research. These ideas are derived from the problems identified above, and are only briefly presented. It will take considerable discussion by the agricultural research community to flesh out the worthiness and practicality of the issues and ideas raised here. Together they can address the issue of establishing a clear and unambiguous picture of agricultural risk and injury in a way that is pertinent, ongoing, systematic, and inclusive. The ideas are not listed in any particular order, and progress in any one area would be an improvement over the present. Certainly, these are possibilities to consider if injury prevention interventions are to ever approach the Healthy People 2010 goals of reducing agricultural work deaths per 100,000 workers by 30%, and nonfatal injury per 100 full-time workers by 30%, respectively, by the year 2010 (US DHHS, 2000).

Chapter Summary

The challenges identified in this chapter are at the core of what keeps production agriculture among the list of most hazardous industries and occupations. Many of the defining characteristics of farming and rural life are one and the same with increased work-related hazard and risk of injury. This adds considerable complexity to accomplishing injury prevention interventions that are effective over time. Program theory and scientific evaluation has been lacking in agriculture safety and health, but there is evidence that this situation is beginning to change. The IER conceptual model and guidebook provide evaluation guides suitable for agriculture. More agricultural hazard and injury data is now available than most professionals can effectively use, yet the insight this data is able to provide to injury causes and remedial factors has progressed relatively little in 20 years. The challenge for agricultural safety and health professionals is to find the ways and means to make a true "paradigm shift" in how to tackle farm safety and health issues. Some ideas on this are presented in Chapters 5 and 6.

Table 4.2. Ideas for improving agricultural injury surveillance and research.

Idea	Comment
Surveillance vs. epidemiological research	The issue of how much and what kind of data can be collected through surveillance activity, in comparison to what should be left for epidemiological studies, needs clarified and communicated extensively throughout the agricultural research community.
National strategy for improving surveillance and research	A national consensus-based strategic plan should be developed to identify priorities, methods, and implementation strategies for surveillance and epidemiological research activity. This strategic plan then needs to be used by agricultural injury researchers at the national, state and community levels.
Risk by source and activity	There is a considerable body of data describing the risk of agricultural injury. Most of this data is presented as a broad person- or place-based rate, such as number of injuries per workers or farms. Broad rates do not provide understanding of the risks associated with major sources of injury, or of the activity involved in the occurrence of the injury. The almost endless combination of injury sources and antecedent activity is likely to make gathering combined exposure-source-activity data difficult, but it is not necessarily impossible. A simpler effort that would still be valuable is to gather exposure-source or exposure-activity.
Categorical data	The agricultural injury research community needs to develop a consensus on a core set of variables for a core set of descriptive categories. At a minimum this should include age groups, sources of injury, and activity at time of injury.
Terminology and definitions	The agricultural injury research and education community needs to develop a consensus on a set of terms that help define the boundaries of terms and combinations of terms such as *work-related, farm work, work environment, occupational agriculture injury*, and the like.
National surveillance of fatal injury	Specific studies are needed to ascertain the ability of the CFOI system to capture cases of fatal injury involving exposure to farm work hazards and environments. Children under the age of 14, volunteer helpers, and work activity interspersed with nonwork activity are areas of particular concern.
National surveillance of nonfatal injury	An ongoing, systematic surveillance system for nonfatal injury is needed. The NIOSH TSIF project was a useful model but needs enhancing so that a rate of injury by exposure to injury sources and work activity at a state level can be developed, and the data published in a timely manner.

5 Human Behavior Theories, Models, and Applications

This chapter explores behavioral theories and models, and their application to agricultural safety and health education. It should be axiomatic that agricultural safety and health professionals must embrace educational approaches in their efforts to reduce agricultural work hazards and risks. The autonomy of farming, cultural beliefs and practices, and ongoing positive reinforcement of risky behaviors points to a heavy reliance upon this approach as a practical reality. Historically, agriculture has relied upon general educational approaches to reduce work hazards, risk and injury. Conceptually, the generic educational approach can be expressed as the "K-A-P" approach, where it is presumed that if people are supplied with sufficient safety and health *knowledge,* they will form or attain the correct safety and health *attitudes* and beliefs, and this will lead to safe and healthful work *practices* (Weiss, 1997).

Much has been written about the relative ineffectiveness of generic educational approaches in comparison to environmental and legislative approaches as a strategy of injury prevention. Problems and relative ineffectiveness issues associated with K-A-P approaches have been noted from many different perspectives, for example, occupational safety and health (Heinrich, 1941; Peterson, 1988; Geller, 1996), public health (Christoffel & Gallagher, 1999; Robertson, 1983) and agricultural safety and health (Aherin et al., 1992; Merchant et al., 1989; Murphy, 1992). The bottom line from these studies and reviews is that the educational approach generally does not lead to permanent desired behavior change. At the same time, several of these authors explain why the educational approach as a methodological strategy of injury prevention should not be abandoned (e.g. Christoffel & Gallagher, 1999; Murphy, 1992; Robertson, 1983). These particular issues will not be additionally reviewed. Specific research-based educational approaches, however, are a different matter.

Within the broad spectrum of agricultural injury prevention are those endeavors that utilize research-based theory and methods to understand and influence the way people think and act as it regards their own safety and health, and of those under their charge, such as children and employees. Broadly based psychological theories, concepts and principles, such as attitude formation, attribution theory, concepts of self, operant conditioning, learning, risk perception and decision-making, have lead to a multitude of more specific theories and principles about why people think and behave as they do regarding personal and occupational safety and health, and methods for modifying those thoughts and behaviors. Many of these theories spring forth from the attempts of applied behavioral psychologists, social psychologists, safety educators, health practitioners, public health advocates, and other like-minded groups, to get people to voluntarily change their safety- and health-related thinking and behavior—thus the connection between broad-based psychological principles and safety and health education.

This literature is generally found under the subject headings of behaviorism, social psychology, cognitive psychology, applied social psychology, applied behavioral analysis, social cognitive psychology, social learning, learning theory, health behavior, health promotion, community-based health promotion, injury epidemiology, and other similar titles. A glossary at the end of this chapter provides a brief description of several terms, concepts, and theories used or referenced in this chapter and the next one. In some instances, items are included because they help explain some other relevant term, concept or theory. Most items defined or discussed in detail in the text are omitted from the glossary.

Modern research-based occupational safety and health theory and methods have evolved from theoretically driven research findings and from experience gained in applied behavioral psychology, cognitive psychology, learning theory, industrial work settings, and in public health disease, injury and disability prevention efforts. To a considerable extent, contemporary research-based industrial occupational safety and health theory and practice has gravitated toward a "behavior-based" approach (Krause, 1997; Geller, 1996). This approach incorporates a relatively straightforward application of psychology and learning principles to workers in the workplace. Public health approaches to occupational safety and health, however, are more ambiguous.

Public health normally approaches all injury issues, including occupational injury, from a population, or epidemiological, perspective. Thus the public health approach to injury prevention is based on surveillance to identify injury problems, identifying risk factors, applying and evaluating interventions, and implementing effective interventions (See Fig 5.1). These four steps are operationalized across the classic epidemiological triad of host, agent, and environment (Christoffel & Gallagher, 1999). Since people (hosts) are one of the three cornerstones of the epidemiological triad, an epidemiological approach to occupational safety and health necessarily includes an interest in human behavior. To the extent that injury research and interventions are directed toward understanding or modifying host factors (human thinking and behaviors), public health relies on the same kinds of psychological and learning principles that industrial safety and health professionals do. Public health professionals are more likely to use the terms "health education," "health promotion" or "psychosocial aspects" (Green & Kreuter, 1991; Cahill, 1996; Christoffel & Gallagher, 1999) to codify the same psychological principles and concepts used in industrial safety and health. Most importantly, professionals working in both industry and public health agree the ultimate goal in occupational safety and health *education* is behavior *change*.

The perspectives of both industry and public health are important to agricultural work injury prevention because: (a) farming is an industrial activity bounded by economic and production issues, (b) the public health approach to occupational safety and health (injury epidemiology) is better suited to addressing the *farm safety—risk paradox* than is a strictly industrial approach, and (c) there is no specific research-based "agricultural" approach to occupational safety and health. This last point does not mean that researchers and educators have never used research-based theory or methods in agricultural safety and health. Indeed, the literature from the 1990s illustrates the initial use of specific theories and models to increase understanding of farm safety behaviors and to formulate interventions. For example, the Theory of Reasoned Action (Aherin, 1987; DeBarr, 1993), the Theory of Planned Behavior (Lee et al., 1997; Petrea, 1996), the Health Belief Model (Gjerde et al., 1991; Hodne et al., 1999; Marlenga & Lee, 1996), the Persuasive Health Message Framework (Witte et al., 1993), Social Cognitive Theory and Social Marketing (Parrot et al., 1996; Parrot et al., 1998) and Narrative Theory (Cole et al., 1997) have all been used to understand and influence farm safety and health related beliefs, attitudes and behaviors.

The farm safety and health literature also show no systematic or comprehensive reviews of the research-based theories and methods from industrial safety or public health that are being applied to agricultural settings. While the actual application of the specific theories and methods shed some light on the appropriateness of various theoretical concepts and models, this is at best a limited or piecemeal approach to understanding theories and models as they may apply to agricultural safety and health. Given the abundance of theories and models that exist for safety- and health-related behavioral change, and considering that none of these have been developed from the perspective of the *farm safety—risk paradox,* a review should be helpful.

Behavioral theory and learning theory are inextricably mixed in occupational safety and health education because safety- and health-related behavior is ultimately shaped and directed by what is learned about hazards and risks. When learning theory is spoken of in the context of occupational safety and health, it is normally in relation to educational strategies, methods and techniques that will influence or direct how people think and behave in relation to the hazards and risks encountered in the workplace or from the work environment. The remainder of this chapter reviews the two foundational approaches to understanding human behavior, looks critically at the safety and health theories and models that flow from these foundational approaches, and then considers these theories and models from the perspective of agricultural safety and health. The following chapter then more explicitly reviews learning theory and connections between it and behavioral models and farm safety and health education.

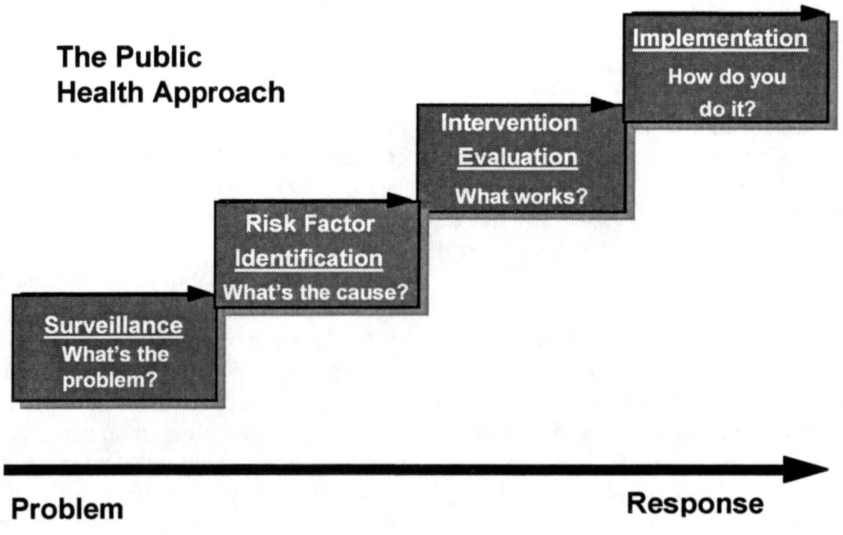

Figure 5.1. The public health approach to injury prevention and control. Source: http://www.cdc.gov/ncipc/pub-res/mr930d.htm.

Behavioral Theory and Applications

Nothing is more confusing than human behavior except for the theories that try to explain it. From the formal beginnings of psychology as a scientific discipline in the late 19th century (Barone et al., 1997; Garnham & Oakhill, 1994), psychologists, philosophers, psychiatrists and other theorists have proposed theories and major schools of thought about the nature, causes and control of human behavior. Two foundational theories—mentalism and behaviorism—form the building blocks for contemporary theories of human behavior (Uttal, 2000). The two foundational theories, in their extreme conceptualizations, point explanations of human behavior in diametrical directions. While few modern psychologists subscribe to the extreme versions of either foundational theory, and in fact promote a combined approach, the continuing evolution of behavioral theories and branches of psychology suggest we have not yet found satisfactory explanations for human behavior within all the situations or circumstances of interest to us.

Scores of textbooks and journals exist within the various areas of psychology. Just a few examples of current texts include: *Handbook of Applied Cognition* (Durso, 1999), *Applied Social Psychology* (Semin & Fiedler, 1996), *Social Cognitive Psychology* Barone et al., 1997), *Situated Learning Perspectives* (Tripp, 1996), *The* Self *in Social Psychology* (Baumeister, 1999), and *Thought and Knowledge* (Halpern, 1996). A few common journal titles include, *Psychological Bulletin, Journal of Personality and Psychology, American Psychologists, Journal of Experimental Analysis of Behavior, Organizational Behavior and Human Decision Making*, and *Psychological Review*. A review of these and similarly titled books and journals reveal the many different ways that the study of human behavior can be organized, researched and practiced. Some topics and methods within the broad psychological literature have a direct relevance to farm safety and health education and behavior change, and it is these topics and methods that are addressed in this review.

These topics and methods of concern are those that relate to what persons exposed to farm-related hazards either do or fail to do, or to what they allow their children or employees to do or to be exposed to. Cognitively oriented theories of human behavior—the modern form of mentalism—dominate today's psychology (Barone et al., 1997; Hunt, 1999; Lee, 1998) and are the types being applied in agricultural settings (identified earlier in this chapter). At the same time, industrial operations throughout the world are achieving significant injury reductions (Krause, 1997) through the application of a form of traditional behaviorism. From the perspective of the *farm safety—risk paradox* (incongruence and disconnect between safety knowledge, attitudes and practices), a question that may be asked is, "Under what circumstances might cognitive and behavioral theories and models of human behavior be applicable to farm safety and health education?"

Before this question can be addressed, a brief review of cognitive and behavioral theories and models and their limitations is necessary. A summary approach to this task is taken because of the vastness of the literature and because many in-depth reviews already exist. After the summary reviews, possible implications to farm safety and health education are discussed.

Cognitivism

The most common behavioral models relevant to farm safety and health education have evolved from cognitive psychologists and health educators' attempts to influence healthy behaviors and lifestyles. Cognitive psychology refers to the study of the mind and how thinking and reasoning influences behavior. Cognitivism coalesced as a distinctive field of study within psychology in the 1950s and 1960s with the advent of digital computers and the metaphor of the mind as an information-processing device (Barone et al., 1997; Garnham & Oakhill, 1994; Halpern, 1996).

From the earliest days of cognitive psychology, there have been two views of man as an information processor: (1) a *literal* view, and (2) a *constructivist* view (Mayer, 1996). In the *literal* view a person performs a series of discrete mental operations on incoming information (input) and uses or stores the product of the mental operations (output). In this view, information can be represented by symbols and be mathematically evaluated. The *constructivist* view is that mental operations represent knowledge rather than information, and this knowledge is constructed in an attempt to integrate incoming experience with existing knowledge. This chapter addresses the literal view of man as an information processor, while Chapter 6 will address the constructivist view.

From a health education perspective, as it became evident that information alone (the K-A-P method) did not sufficiently motive people to change or modify unhealthy behavior to healthy behavior, health educators turned to psychological theory and concepts to understand why (Allen, 1998). Literal view theories and concepts applicable to health education and promotion are numerous and are organized and discussed under an array of terms. For example, two popular theories of health behavior, the Health Belief Model (HBM) and the Theory of Reasoned Action (TRA), have been reviewed under the broad banners of health psychology (Allen, 1998), social cognition models (Connor & Norman, 1996), theories of health-protective behavior (Weinstein, 1993), expectancy-value theories (DeJoy, 1996), and continuum (stage) theories (Weinstein et. al., 1998).

Regardless of how theories and models are named or grouped, most have several common features because they all have evolved from the same conceptual approach to human behavior. Several authors, including Lee (1998), Uttal (2000) and Weinstein (1993) have noted most theories of human behavior, developed since the advent of

Table 5.1. Behavioral and health behavior theories and models.

Major Sources	Theory or Model	Major Purpose, Prediction	Assumptions, Foundational Statements	Major Dimensions, Constructs	Major Criticisms, Flaws
Rosenstock, 1966; Rosenstock, 1974; Janz & Becker, 1984; Sheeran & Abraham, 1996.	Health Belief Model (HBM)	Purpose: Identification of appropriate targets (health beliefs) for health education.	1. Value is placed on avoiding illness. 2. A specific action will prevent (ameliorate) illness. 3. Some cue was needed to trigger decision-making. 4. Other variables might indirectly influence health behavior.	Perceived: (a) susceptibility, (b) severity, (c) benefits, (d) barriers.	1. How variables combine to produce behavior is not explicitly defined. 2. Variables such as intention, social pressure, self-efficacy are not considered. 3. Considerable variance in how variables are defined and operationalized in studies. 4. Ignores cue variables. 5. Often considered a short list of variables rather than a theory.
Rotter, 1954; Rotter, 1966 Wallston et al., 1978; Norman & Bennett, 1996.	Health Locus of Control (HLC)	Purpose: To predict future preventive health behavior. Prediction: Persons with internal HLC more likely to engage in health-promoting behavior.	1. Likelihood of behavior is a joint function of: (a) expectancy that behavior will lead to a particular reinforcement, (b) the extent to which the reinforcement is valued. 2. Control of events is seen as internal (one's own actions) or external (environment, actions of others). 3. Persons with internal locus of control believe events are under personal control. 4. Persons with external locus of control believe events are determined by factors beyond their control.	The extent to which people believe: (a) health is result of their own actions (internal), (b) health is under control of powerful others (external), (c) health is due to chance or fate (chance).	1. Value people place on health not accounted for in most applications. 2. When health value is considered results are mixed. 3. Measurement scales are not situation or behavior specific. 4. Having high external control and chance control beliefs are adaptive in some circumstances. 5. Generally obtains weak predictions; too narrow to explain health behavior.
Rodgers, 1975; Rodgers, 1983; Boer & Seydel, 1996.	Protection Motivation Theory (PMT)	Purpose: Initially to provide conceptual clarity to understanding of fear appeals; revised to extend to general theory of persuasive communications. Prediction: Adaptive response to health threat increased if person feels (a) vulnerable, (b) threat is severe, (c) can respond effectively.	1. Fear initiates a corresponding cognitive mediating process. 2. Adaptive and maladaptive coping with health threats results in two processes: threat appraisal and coping appraisal. 3. Operationalized as intention to perform health-protective behavior or avoid health-compromising behavior. 4. Actual behavior assumed to be a function of intentions. 5. Behavior reinforcement occurs if recommend action is successful (threat is reduced).	1. Severity — how severe are the consequences? 2. Vulnerability (susceptibility) — how likely is threat to actually happen? 3. Response efficacy — how effective is recommended protective behavior? 4. Self-efficacy — am I able to perform the recommended behavior? 5. Protection motivation — do I intend to perform the recommended behavior? 6. Protective behavior — performing the behavior?	1. Threat appraisal components result in weak predictions. 2. Lack of consistency in conceptions and operationalizations detract from explanatory power.
Fishbein & Ajzen, 1975; Ajzen & Fishbein, 1980; Ajzen, 1985; Ajzen & Madden, 1986; Ajzen, 1991; Conner & Sparks, 1996.	Theory of Reason Action (TRA); Theory of Planned Behavior (TPB)[a]	Purpose: To predict and explain human behavior in specific contexts. Predictions: (a) that behavioral intentions are the immediate antecedents to (causes of) behavior; (b) course of action will be followed if person believes	1. Behavioral decisions are based on careful (rational) consideration of available information (alternatives). 2. Aggregate behavior better represents underlying behavioral disposition than any single behavior. 3. Intentions capture the motivational factors that influence a behavior. 4. Intentions are determined by attitudes toward behavior and subjective norms.	1. Determinants of intention: attitudes, subjective norms, perceived behavioral control. 2. Determinants of attitudes: salient beliefs about consequences of behavior based on likelihood of behavior occurring (outcome) and evaluation of that outcome. Issue of specificity is critical. 3. Determinants of subjective norms: normative beliefs, which are perceptions	1. Doesn't account for influence of past behavior. 2. Does not include concept of health threat. 3. May be too elaborate to be practical. For example, each behavioral alternative, and different frequencies of repeated behavior, should be assessed. 4. Measurement of some constructs not always adequate (low internal

		behavior (action) will lead to outcome s/he values, if significant others value the behavior, have necessary opportunities and resources to carry out behavior.	5. Attitudes and subjective norms are explained by beliefs about consequences of performing behavior and expectations of relevant others. 6. Intentions and behavior can be modified by lack of opportunity and/or resources (volitional control).	of significant others to perform or not perform the behavior. 4. Determinants of volitional control: individual's perceptions of whether performance of the behavior is easy or difficult (possible or impossible).	reliability). 5. Model addresses perceptions of control rather than actual control.
Bandura, 1977; Bandura, 1986; Schwarzer & Fuchs, 1996; Barone et al., 1997.	Self-Efficacy (S-E)	Purpose: To account for judgments of one's ability to perform specific behaviors. Prediction: Higher self-efficacy will lead to higher goals, better solutions, and greater achievements.	1. Behavioral change is facilitated by a personal sense of control. 2. Self-efficacy is closely tied to goal settings and self-motivation. 3. Self-efficacy influences: (a) goals people set for themselves, (b) choice, effort and persistence of goal-directed activity, (c) efficiency and effectiveness of problem solving and decision-making, (d) selections of situations and activities to become involved in.	1. Optimistic self-beliefs. 2. Magnitude, generality and strength of outcome expectancies. 3. Sources of efficacy expectations: (a) performance accomplishments, (b) vicarious experience, (c) verbal persuasion, (d) emotional arousal.	1. Lacks explicit consideration of social influences. 2. Fails to incorporate measures of value associated with different outcomes.
Prochaska & DiClemente, 1983; Prochaska, DiClemente & Norcross, 1992; Prochaska et al., 1994; Sutton, 1997.	Trans-theoretical Model [b] (TTM)	Purpose: An integrative and comprehensive model for describing when and how problem behaviors change.	1. People intentionally change behavior with and without psychotherapy. 2. Key question: Are there basic, common principles that can reveal the structure of change occurring with and without psychotherapy? 3. Research shows: (a) a cyclical pattern of movement through specific stages of change, (b) a common set of processes of change, (c) a systematic integration of the stages and processes of change.	Stages of change: 1. Precontemplation. 2. Contemplation. 3. Preparation. 4. Action. 5. Maintenance. Processes of change: 1. Consciousness raising. 2. Self-reevaluation. 3. Self-liberation. 4. Counter-conditioning. 5. Stimulus control. 6. Reinforcement management. 7. Helping relationships. 8. Dramatic relief. 9. Environmental evaluation. 10. Social liberation.	1. Does not describe "how" people change. 2. Assumptions about causal pathways are not made explicit. 3. Model not strongly supported by empirical longitudinal studies.
Locke & Latham, 1990; Locke, 1991; Barone et al., 1997.	Goal Theory	Purpose: To specify the factors that affect goals and their relationship to action and performance. Prediction: Specific and difficult goals result in better performance.	1. Goals are immediate regulators of human action. 2. Goals direct attention and action. 3. Concept of motivation explains direction, effort and persistence of goals. 3. Four mechanisms by which goals affect task performance: (a) directing attention and action, (b) mobilizing energy or effort, (c) prolonging effort over time, (d) development of strategies (creative problem solving) for goal attainment. 4. Past performance predicts future goals. 5. Goals specify the conditions for positive and negative self-evaluations.	Issues affecting goal performance: (a) goal range and type, (b) goal specificity, (c) personal ability, (d) knowledge of results (feedback), (e) monetary rewards, (f) participation and supportedness, (g) individual differences, (h) goal acceptance and choice, (i) type of and control over tasks.	1. Most research is based on simple tasks in which applying more effort produces better results. 2. Most studies assign goals to subjects, masking differences between subjects. 3. Most difference variables lack theoretical rationale making results difficult to interpret. 4. Inconsistency among personality measurement variables.

[a] TPB is a revision and extension of TRA (Ajzen, 1985).
[b] The contemporary research literature includes variations of these theories and models as researchers attempt to improve identified weaknesses and extend them to new areas of interest.

cognitive psychology, are relatively minor deviations of an existing cognitive theory that the new models are trying to improve upon. Table 5.1 is a summary compilation of literal view cognitively oriented theories and models commonly used to understand, predict and change health-related behaviors (Conner & Norman, 1996; Elder et al., 1998; Weinstein, 1993). Several have been applied to agricultural safety and health. The table includes pertinent information about the theories and models, including common literature sources, origins, development, assumptions, etc. The table's purpose is to provide enough information to help convey the essence and major characteristics of the model or theory without going into an extensive review of it, since this already has been done in the literature. Readers wanting a more in-depth understanding of these models can consult the citations listed in the table, or in this and the preceding paragraphs. The book, *Predicting Health Behaviors,* edited by Connor and Norman (1996), was relied upon heavily to develop Table 5.1.

Table 5.2 summarizes how some health educators and researchers have cited similar and overlapping constructs among literal view theories and models of human and health behavior (See Table 5.2 notes). Underlying most of these theories are the family of "expectancy-value" theories (Barone et al., 1997; Strobe & de Wit, 1996). Expectancy-value theory, which predates cognitive psychology, was developed by Kurt Lewin (1890-1947), J.B. Rotter (1954) and others, and posits that behavior can be predicted from expectancies (beliefs) and values (goals, needs), rather than associations and reinforcement as advanced by behaviorists (Barone et al., 1997; Strobe & de Wit, 1996). Expectancy-value theories focus on changes in expectancies, values and goals of people, and how those expectancies, values and goals may be changed, rather than on behavior change itself.

A familiarity with the ideas expressed in Tables 5.1 and 5.2 will help readers understand literal view cognitive approaches to behavior change. The last column in Table 5.1 identifies some of the general criticisms associated with each theory or model. An important point to keep in mind is that many of the criticisms identified with a theory come first from the developer or creator of the theory. Psychological theorists acknowledge that no one theory or model can apply to all human thinking or behavior in all situations. The relevance of these general criticisms to farm safety and health education will be discussed in more detail after the section on behavioral approaches to understanding and controlling human behavior.

Behaviorism

The behaviorism approach to occupational safety and health is embodied in far fewer theories and concepts than is the cognitive approach, and these are explained below. The scientific application of behavioral theory to occupational safety and health can be described under the single banner of Behavior-Based Safety (BBS). The agricultural safety and health literature is nearly devoid of discussion of behaviorism and BBS. This chapter helps to fill this void, but readers should not interpret the space or detail given to behaviorism and BBS as more than an attempt to provide an adequate explanation of its development, constructs, and methods as they may be applicable to general occupational and occupational agriculture safety and health. A summary explanation of behaviorism is provided first, followed by a more detailed explanation of the BBS process. Behaviorism as a theory of behavior is generally credited to the early 20th century work of John B. Watson (1878-1958) (Barone et al., 1997; Uttal, 2000), who systematized the doctrines of behaviorism and popularized them among psychologists (Garnham & Oakhill, 1994). Watson was strongly influenced by Edward L. Thorndike (1874-1949), an experimental psychologist who was interested primarily in learning theory. Thorndike's research with laboratory animals resulted in the "law of effect" as a major theory of learning in educational psychology. The "law of effect" had two major postulates (Garnham & Oakhill, 1994). The first was that if an animal acts in a particular way in a given situation, and if the outcome is positive, the same action is more likely in similar situations in the future. The second was that learning typically occurred by trial and error.

Watson advocated that experimental psychological theory and research should focus on observable aspects of behavior. Additionally, he believed the laws of learning applied to both humans and animals, but that animals were more convenient to study. Watson's behaviorism did not reject the notion of thinking or mental processes (cognitivism) but he felt, in essence, that the mind was not accessible for study. Watson suggested that behaviorists must use a scientific vocabulary that did not include quasi-subjective terms such as sensation, perception, image, desire, emotion, etc. (Uttal, 2000).

Although Watson was the founder of behaviorism as a science, the name B.F. Skinner (1904-1990) is a more recognized name associated with behaviorism in American contemporary society (Sternberg, 1999). His work is still referenced in textbooks and journal articles, and in particular among behavioral psychologists that advocate behavioral approaches to occupational safety and health. For example, Geller (1996), in *The Psychology of Safety*, refers to Skinner as the "the founder of behavioral science and its many practical applications, ..." (p. 11).

Skinner's behaviorism has been referred to as "radical behaviorism" in the sense that he carried the principles and tenets of Watson's behaviorism to an extreme (Uttal, 2000; Lee, 1998; Barone et al., 1997). Although he did not reject the reality of mental states, he more completely rejected the idea that mental states had any usefulness in studying the shaping of human behavior (Skinner, 1974). His faith was unshakable that the results of tightly controlled, empirical-based laboratory research with animals could be translated to both an explanation and shaping of human behavior, without any loss of predictive power. Skinner advanced the argument that most of human life, including learning, problem solving, and cultural practices, was explainable in terms of operant conditioning (Proctor & Weeks, 1990;

Table 5.2. Similarities in cognitive theories and models of human behavior and health behavior.

Source	Theories and Models	Similarities
Weinstein, 1993; Weinstein et al., 1998	1. Health Belief Model (HBM) 2. Subjective Expected Utility (SEU) Theory 3. Protection Motivation Theory (PMT) 4. Theory of Reasoned Action (TRA) and Theory of Planned Behavior (TPB)	1. Assume anticipation of negative health outcome and desire to avoid or reduce outcome creates motivation for self-protection. 2. Impact of negative outcome on motivation to act depends on beliefs about likelihood of negative outcome. 3. Questions to assess major constructs are essentially indistinguishable from one theory to another. 4. Assume expected benefits in risk reduction weighed against costs of acting to predict changes in behavior. 5. Do not predict the amount of precautionary behavior that will occur; only the relative likelihood of action is predicted. 6. Variables that influence action are combined to predict behavior in a single prediction equation.
Elder et al., 1998	1. Health Belief Model (HBM) 2. Theory of Reason Action (TRA) 3. Stages of Change (SOC) (e.g. TTM)	Person must: 1. Have a strong positive intention or predisposition to perform a behavior. 2. Face a minimum of environmental barriers to performing the behavior. 3. Have the requisite skills for the behavior. 4. Believe that positive outcomes will follow performance of the behavior. 5. Believe that there is normative pressure to perform the behavior and none sanctioning the behavior. 6. Believe that behavior is consistent with the person's self-image. 7. Have a positive effect regarding the behavior. 8. Encounter cues to act at an appropriate time and place.
DeJoy, 1996	1. Health Belief Model (HBM) 2. Theory of Reason Action (TRA) and Theory of Planned Behavior (TPB) 3. Protection Motivation Theory (PMT)	Constructs relevant to workplace self-protective behavior: 1. Threat-related beliefs, 2. Self-efficacy, 3. Response efficacy, 4. Barriers, 5. Normative expectations.
Norman and Conner, 1996	1. Health Belief Model (HBM) 2. Health Locus of Control (HLC) 3. Protection Motivation Theory (PMT) 4. Theory of Reasoned Action (TRA) and Theory of Planned Behavior (TPB) 5. Self-Efficacy (S-E)	Theoretical comparisons: 1. Focus on the notion of health threat as measured by perceived susceptibility and perceived severity. 2. Focus on perceived consequences of performing a health behavior. 3. Focus on importance of perceptions of behavioral control or barriers. 4. Normative influences on behavior not explicitly covered, with exception of TPB. 5. TPB and PMT include an intervening variable that is seen to mediate the relationship between other social cognitive variables and behavior (e.g. behavioral intention).

Note 1: Some specific comments by authors may not apply to all theories and models listed. See citation(s) for specifics.
Note 2: Some reviews include theories and models not included in Table 5.1.

Skinner, 1974). Operant conditioning refers to behavior that is followed by a consequence: the nature of the consequence determines or modifies the tendency to repeat or not repeat the behavior in the future.

Key concepts in Skinner's behaviorism relevant to this review include (Skinner, 1974):

- The human body has three nervous systems: the *interoceptive* system, which carries stimulation from organs like the bladder, glands, blood vessels, etc., and operates for the internal efficiency of the body; the *proprioceptive* system, which carries stimulation from the muscles, joints, tendons, etc., of the skeletal frame and other organs involved in the maintenance of posture and movement; and the *exteroceptive* system, which is concerned with seeing, hearing, tasting, touch, etc., in the world around us.
- Humans act to achieve pleasure or satisfaction and to escape from or to avoid pain and discomfort (hedonism).
- Many human behaviors have "survival" characteristics about them (e.g. eating food, drinking water). From this, it follows that some behavior is strengthened by its consequences, and for that reason the consequences themselves are called "reinforcers."
- Behavior that is reinforced by a consequence is more likely to occur again. A positive reinforcer strengthens any behavior that produces it (e.g. a glass of water is reinforcing when we are thirsty). A negative reinforcer strengthens any behavior that reduces or terminates it (e.g., removing a shoe that is pinching your toes).
- An organism (animals, humans) learns to react discriminately to the world around it under certain contingencies (occurrences) of reinforcement. The probability that an organism will respond in a given way changes as contingencies of reinforcement change.
- The genetic endowment of the species (in this case human), plus the contingencies to which the individual has been exposed, determines what a person perceives. We are not merely "mindful" of the world about us; we respond to it in idiosyncratic ways because of what has happened when we have been in contact with it.
- Many contingencies involving private (mental) stimuli follow from simple mechanical relations among stimuli,

responses and reinforcing consequences. The organism may or may not be consciously aware of the stimuli.
- We do not need to describe contingencies of reinforcement to be affected by them.
- Thinking (mentalism, cognitivism) is "covert behavior," or "behavior executed on a small scale" (not observable).
- Talking (speaking) is verbal operant behavior. It has reinforcing and contingency characteristics (e.g. positive, strengthening, immediacy).
- All behavior, including mental behavior such as thinking, perception, knowledge, attention, feelings, emotions, motivation, etc., is attributable to the subtle and complex relations among three things: the situation in which behavior occurs, the behavior itself, and its consequences.
- Mentalism directs researcher's attention away from observable antecedent events and toward the unobservable that can never be as objective or reliable as observable events.
- Mental states cannot be seen, therefore they cannot be observed or exactly described, so interpretations of mental states can at best only describe behavior, as opposed to explaining it.
- By himself, an individual can acquire very little behavior with respect to the future in his own lifetime, but as a member of a group he profits from the social environment (culture) maintained by the group.

Skinner's version of behaviorism is important because the contemporary industrial approach to behavioral safety is based on his theories about operant conditioning. The relevance of behavioral approaches to occupational safety and health has been articulated from at least the time of Heinrich (1941) and became more numerous in the 1970s (e.g., see Fitch et. al., 1976; Grimaldi, 1970; Peterson, 1975; Smith et al., 1978). The pioneering of a systematic application of applied behavioral analysis methods to industrial, community and public safety and health issues is generally credited to the 1970s and 1980s research and work of behavioral psychologists such as Judith Komaki, Beth Sulzer-Azaroff, Thomas R. Krause and E. Scott Geller. Current major textbooks on this subject include those by Krause (1995, 1997), McSween (1995), Geller (1996) and Peterson (1998). Recent major journal article reviews and meta-analyses of specific applications of the behavior-based process include those by DePasquale and Geller (1999), Krause et al. (1999), and Sulzer-Azaroff and Austin (2000).

Krause and associates in 1980 were the first to systematically combine applications of behavioral science principles, behavioral management theory, and statistical methodology and apply them to industrial work settings on the behalf of occupational safety (Krause et al., 1990). They subsequently named their applications the "behavior-based safety process." Today their process is more widely referred to as "behavior-based safety" (BBS) (Krause, 1997).

The BBS Process

Behavior-based safety, as a theoretically-driven process, has gained considerable support from many industrial safety professionals in recent years. Presumably, this is because of the documented successes it is achieving across a wide spectrum of industrial applications. For example, in the reviews by Sulzer-Azaroff and Austin (2000) and Krause et al., (1999), 155 different companies are cited as documenting significant reductions in injuries and accidents. All companies had implemented some type of BBS system for between one and five years, and represented companies with from five to 3,000 employees. Guastello (1993) found BBS (seven sites) to be the most effective approach in his analysis of 10 different approaches to accident prevention programs.

Although there are many variants of the BBS process, not all are successful because of confusion about what it really means to implement one (Veazie, 1999; Blair, 1999; Peterson, 1997). In particular, BBS is not "behavior modification" (Sulzer-Azaroff & Lischeid, 1999). The essential components and principles of the behavioral science approach to BBS are explained below. The writings of Krause (1995; 1997), Krause et al. (1999), Geller (1996) and McSween (1995) are relied upon heavily for this explanation. All three are applied behavioral science practitioners associated with consulting firms. Hypothetical examples from agriculture are used to help explain some important points.

True behavior-based safety relies on a combination of research findings and methods of applied behavioral analysis, Total Quality Management (TQM), and organizational behavior management. The influence of W. Edwards Deming (1900-1993) and his TQM methods for improving quality and production systems are significant. Similarities between the two fields include:
- Operational definitions of those variables known to be critical for excellence.
- Measurement of targeted variables.
- The display of data to assist in understanding variation.
- Feedback as a learning mechanism.
- Use of data to quantify, measure and provide information enabling workers to determine relationships between variables.
- Problem solving as an ongoing contribution to continuous improvement.
- The role of manager is to initiate and support process improvements.
- The role of worker is to use the method actively in daily activities.

According to Krause (1997), "In the same way that scientific research methodology is applied to improve quality (operational definitions, measurement, feedback), the behavioral approach to performance applies the same methodology to a variety of behavioral issues—in this case safety performance" (p. 29).

Each application of the BBS safety process is unique because it depends on a structured process to identify and correct specific at-risk behaviors. A key to this structured process is that workers, who are doing the work at the specific site where behaviors are being observed and corrected, drive the process. While there is considerable flexibility in the structured process because of the considerable range of workers, work, worksites and organization and business management schemes that exist, the BBS process can be explained by three fundamental concepts that drive the process. Krause (1997), Geller (1996) and McSween (1995) have different ways of approaching these concepts, but the concepts are nevertheless at the heart of their applications of the BBS process. These three concepts are: (1) a critical-mass approach to defining and measuring at-risk behaviors; (2) understanding the relationship of antecedents, behaviors, and consequences (ABC); and (3) a management-supported, worker-driven safety system. Each of these fundamental concepts is briefly reviewed below.

Critical Mass of At-Risk Behaviors

There are several points to understand within this concept. The idea behind the word "critical" is that a specific behavior is the central factor in most work injury incidents. Identifying a specific behavior within jobs or tasks that leads to work injury avoidance (safe behavior) or to injury (at-risk behavior), is seen as paramount importance, hence use of the word critical. "Critical mass" means accumulating (by identifying and recording through observation) the mass of safe and at-risk behaviors critical to specific job or task performance across the facility or workplace. This is normally done over several weeks or perhaps months, depending on the size of the workplace and frequency of jobs and tasks. Safe and at-risk behaviors are specifically identified and described by reviewing job hazard analyses, observing work practices, interviewing workers and supervisors, and reviewing injury records. A key factor in identifying a critical behavior is that behavior is defined as "observable acts or actions." This leads to a focus on what a worker does or does not do, rather than on attitudes, feelings or emotional states, and to the "outcome" of a specific behavior.

In BBS, behaviors are seen as either "at-risk" or "safe." One major purpose (other purposes will be discussed later) in recording these behaviors is to compute their percent of occurrence within the total set of behaviors for a specific job or task. For example, a task such as hitching a wagon to a tractor may be broken into five behavioral steps. Each step can be observed and recorded as at-risk or safe. If two behaviors were observed as violating the description of how they should be safely done, 40% (2 of 5) of the behaviors would be identified as at-risk and 60% (3 of 5) as safe behaviors. Hitching a wagon to a tractor, of course, may be only one of several tasks in a given day that could be observed. The idea is to make as many observations on the same task for as many workers and for as many tasks as needed to improve safety performance. The goal is always to strive for 100% safe behaviors for all tasks for all workers.

An analogy of product sampling for quality control may be useful in understanding the concept of critical-mass behavior. A defective product at the end of a production line, for example, a badly bruised apple in a shipping carton, is an indication that something needs to be adjusted in the sorting or inspection lines. In the same way, a work injury, such as a strained back, means something needs to be adjusted in the way work is being accomplished. The bruised apple and the strained back both represent "after the fact" observations of something that needs adjusting or fixing. A standard technique for improving quality control is to sample products *upstream* before they result in a defective product. In the case of the apple, a sorter may need adjusting to better find and remove bruised apples. In the case of the strained back, lifting, turning or carrying motions may need to be adjusted before the result of strained back muscles. Just as sampling products for quality is done systematically and at stages before a final product is produced, critical behavior should be sampled systematically and at stages before an injury results.

A focus on outcomes of behavior in safety leads to identification of human behavior as a common final link and pathway to an injury incident in most cases. But in BBS, this does not mean a worker is "blamed" for the incident. Behavior-based safety advocates are clear that blaming workers for accidents is not a part their science (Sulzer-Azaroff & Lischeid, 1999) and is counter-productive to a continual safety improvement process. At the same time, success at identifying critical behaviors is crucial to the concept of BBS. Geller (1996) advocates a technique for identifying critical behaviors with the acronym SOON, which means, be:

- Specific (concise behavioral definition, unambiguous language).
- Observable (overt behaviors, countable and recordable).
- Objective (no interpretations or attributions, record "what," not "why").
- Naturalistic (normal interaction, real-world activity).

To help explain the connection between a critical mass of at-risk behavior and injuries, Krause (1997) uses an analogy of the critical mass of a radioactive substance and its explosiveness:

> For instance, in the case of a piece of uranium of critical mass, no one has any idea precisely which unstable atom is going to touch off the chain reaction resulting in an explosion. Just as there is randomness or unpredictability about individual atoms in a mass of radioactive uranium, so there is a randomness and unpredictability about individual employee behaviors at a particular facility. On the other hand, the activity of the whole mass of uranium is *statistically* very predictable. And in the same way, the overall safety performance of an entire facility is statistically very predictable; at a given level of at-risk behaviors there will be explosive events—accidents are going to follow (p. 31).

Past accident and injury incident investigations make it clear that a large number of at-risk behaviors and/or conditions precede almost every incident. It is impossible to know *a priori* which specific act will, so to speak, be the act that "breaks the camel's back." The BBS process minimizes the chance of that last straw by: (a) operationally defining safe behaviors for specific work at specific locations, (b) making observations about safe and at-risk behaviors, and (c) using these observations to correct at-risk behavior. In fact, the process can prevent the load on the camel from even becoming heavy. Furthermore, the BBS process, through these three specific steps, seeks to identify management systems, physical plant conditions, production processes and other such factors that implicitly or explicitly condone or encourage at-risk behaviors. The concept of identifying and using critical behaviors is closely connected to the concept of a worker-driven safety system and will be discussed further in that section.

Antecedents, Behaviors, and Consequences (ABC)

A basic premise of BBS includes three primary points regarding causes and effects of behavior:

1. Behavior is a function of the immediate environment;
2. Events that occur in the immediate environment can be divided into those that occur before the behavior (antecedent or trigger event) and after the behavior (consequence); and
3. Both antecedents and consequences influence behavior, with consequences a stronger and more reliable predictor of behavior than antecedents (McSween, 1995). Implicit in this premise is that behavior is goal-directed, or people act in ways to accomplish some goal or end result (Locke, 1991; Barone et al., 1997). The establishment of this relationship between behavior and before/after events is generally credited to the work of Skinner and his experimental learning work with mice, rats, pigeons and other laboratory animals. A key finding by Skinner was that voluntary behavior can be strengthen (increased) by a positive consequence received immediately after the behavior, while a negative consequence received would weaken (lessen) the behavior. Skinner named his research "operant conditioning" because the research subjects (animals) learned to *operate* on their environment to obtain a certain consequence (Skinner, 1974).

Antecedents (triggers) consist of both internal and external factors. Internal factors consist of attitudes, beliefs, values, motivations, peer pressure and other mental (cognitive) states. External factors consist of physical conditions, objects and facilities. Of internal factors, safety attitudes cause most of the confusion regarding motivations for at-risk and safe behavior. Behavior-based safety advocates recognize that attitudes and behavior are a two-way street, that is, attitudes can influence behavior and behavior can influence attitudes (e.g. see Geller, 1996). The BBS process works on the *behavior influences attitudes* side of the street, even while recognizing the street goes both ways. The behavior-attitude side is favored because behavior can be observed and managed much more readily than can attitudes (or any other internal state). Because attitudes are internal events that cannot be observed, measured or managed on a daily basis, it is not seen as an efficient method for improving safety performance.

External antecedents are conditions or situations that are a part of the immediate environment where the behavior is occurring, and that play a role in triggering behavior. For example, if a room is dark and a person wants to see, the "darkness" is the antecedent to turning on the light switch. If a person is chilly, the coldness is the trigger to putting on a sweater. On worksites, the condition and arrangement of facilities, work tools and products often act as a partial cause (antecedent) of a specific behavior.

Behavioral consequences can be *immediate* or *delayed*, *certain* to occur or have *uncertainty* about them, and be viewed as *positive* or *negative* by the person about to commit the behavior. Krause (1997) has labeled these the "soon-certain-positive" features of consequences. Table 5.3 shows the possible combinations of the features of consequences. Behavioral science research has shown that consequences having all three features are more powerful than consequences that are delayed, uncertain, and negative, or some combination of these. The direct connection between these features and occupational safety and health is that at-risk behavior normally results in a soon, certain and positive consequence for the individual. Some illustrations and examples can help clarify the relationships between antecedents, behaviors and consequences.

Table 5.3 illustrates the common relationship between antecedents, behaviors and consequences. A ringing telephone can be used to illustrate the A-B-C relationship more vividly. If a telephone rings (an antecedent event) a person normally answers it (the behavior) and talks to someone (the consequence) on the other end of the line. In this scenario it seems that the antecedent event (phone ringing) influences the behavior (answering the phone) and in fact it does. And as long as someone is always on the other end of the line when it rings, the antecedent (phone ringing) will strongly influence the behavior (answering the phone) because the consequence (talking to someone) is always soon (instantaneous), certain (someone's always been there before), and positive (expectation has been met that someone is on the other end of the line). But what if no one was on the other end of the line? How many times would a person answer a phone if no one was on the other end of the line to talk to? Eventually, the consequence of no one being at the other end of the line to talk to would result in the person not picking up the phone when it rang. Undoubt-

Table 5.3. The range of consequences.

STRONGEST	soon	certain	positive
STRONGER	not soon	certain	positive
	soon	not certain	positive
	soon	certain	not positive
WEAKER	not soon	not certain	positive
	soon	not certain	not positive
	soon	certain	not positive
WEAKEST	not soon	not certain	not positive

edly, there are many examples in everyday life where simple little consequences change behavior.

The issue of positive and negative consequences may need further explanation in the context of goals and at-risk and safe behaviors. Research strongly supports the idea that most human action is motivated by the desire to reach some result or end state—a goal (Barone et al., 1997; Bandura, 1977; Locke, 1991). Goals may involve some external state such as a physical object or a skill, or they may involve an internal state such as knowledge, happiness or satisfaction. People may consciously think about their goals, or may go about their days barely conscious of them. The less conscious they are of their goals and the behavior they direct, the more likely is the behavior to be labeled "habit" (Manstead, 1996). A widely accepted premise in behavioral psychology is that much of everyday behavior is by habit. Another widely held premise, first established by Clark L. Hull (1884-1952) through animal laboratory experiments, is that when a response can be made to achieve a goal, if the amount of work is too high, the organism may not make the response or choose an alternate response that requires less effort to reach the goal. This experimental finding suggests organisms (e.g., humans) tend to exhibit responses that allow them to reach their goals with the least amount of work or effort. More simply, when faced with a task they wish to accomplish (goal), humans often will select the quickest and easiest way to complete the task to save time and effort (O'Donohue & Kitchener, 1999). In the context of work safety, this premise suggests humans often will attempt to fulfill their work goals with the least amount of time, effort, cost and inconvenience to themselves. This practice is commonly referred to as taking "shortcuts." The use of risk-taking shortcuts as factors in occupational injury events is well documented in injury and fatality investigations.

At-risk behaviors are generally more complex and of a substantial nature than a ringing telephone. For example, consider the issue of a 15-year-old farm boy, who is to go out by himself to bale hay for the first time. The field where the baling will be done is not within sight of the farmstead or a roadway. The baling was expected to take four to five hours to complete (the goal). The boy was instructed on the safe operation of the hay baler by his father, had observed his father and neighbors operate hay balers for many years, and had participated in several farm safety activities and discussions in 4-H and agricultural science classes in high school. His father, as the boy was leaving for the field, told him to be careful and to take his time. As luck would have it, after about three hours of baling, hay started to bunch up at the intake chamber where the bale is formed and stopped feeding into the chamber. Almost instantly the boy recognizes he has to get off the tractor and do something about the plugging up of the hay. What antecedents and consequences might influence the boy's behavior of checking and clearing the bale chamber? Table 5.4 illustrates how goals, antecedents, behaviors and consequences may interact.

If the boy follows his dad's instructions, and perhaps even his own instincts, and shuts off the tractor, then the consequence is to struggle considerably more to clear the plug and be delayed longer getting back to baling. In the few seconds he takes to think about his choices, shutting off the tractor does not come across as a positive thing. Because he has seen his dad and others do it this way many times, he knows that leaving the tractor run will save him time, effort and aggravation. This he views as a positive consequence. What about the possibility of injury? How does that come into play? The boy has already learned not all at-risk behaviors or chance-taking results in a negative consequence, such as a serious injury, and risky behavior is usually rewarded by saving time and effort. He has learned this through his own experiences in work, recreation and everyday life, and through observing others in the same venues. A serious injury is a relatively rare event for any single person. For example, in 1999 the incident rate for nonfatal occupational injury treated by hospital emergency rooms was 2.8 per 100 full-time workers per year (NIOSH, 2000, p. 89). This is another way of saying that in any given year, the injury incidence rate is one injury per 35.7 person-years of work (Krause, 1995). The threat of an injury, even possibly a life-threatening injury, does not normally override the value of immediate and relatively easy and certain goal attainment.

One goal of the BBS safety management process is to make safe behaviors more attractive and at-risk behaviors less attractive. This goal can be operationalized through manipulating antecedents and consequences. In an earlier paragraph it was explained that the BBS process prefers to work on the behavior-attitude side of the street rather than the attitude-behavior side of the street. Another way of making this same point is to say that the BBS process prefers to work with behavior-consequence relationships rather than with antecedent-behavior relationships. Four major strategies for working with behavior-consequence relationships are to:

1. Increase positive consequences of safe behavior,
2. Decrease negative consequences of safe behavior,
3. Decrease positive consequences of at-risk behaviors, and
4. Increase negative consequences of at-risk behavior.

The same strategy, of course, applies to antecedent-behavior relationships.

The benefit of mapping out behaviors in an ABC context is to identify opportunities to intervene to change consequences and antecedents. There are many ways to do this, both traditional and nontraditional. The BBS safety process incorporates the behavioral principle of *feedback* to increase positive consequences of safe behaviors, decrease the positive consequences of at-risk behavior and to create a supportive safety culture within the organization. The behavioral feedback principle and a supportive safety culture are operationalized through a management supported worker-driven safety system—the third key concept of the BBS process.

Table 5.4. Interaction of goal, antecedents, behavior and consequences.

Possible antecedents	Behavioral choice	Possible consequences					
		Soon	Delayed	Certain	Uncertain	Positive	Negative
External factors A hot and dry day, considerable dust is in the air. Tractor is old and runs roughly and loudly, is without a cab or umbrella. He has no hearing protection. Can see dark clouds beginning to gather in the distance. ***Internal factors*** Irritable because he's hot, dirty, weary, and has a headache. Feels pressure to get back to baling because it might rain soon. Remembers being told to shut off the tractor in such situations but remembers seeing his Dad and neighbors clear similar plugs with the power to the baler still running. Remembers his Dad shutting off the baler one time to clear a plug, and how his Dad cussed about how long it took and how hard it was. Remembers his friends making fun of him a few months ago because he had a perfect score on a farm safety quiz in school.							
	Shut off tractor before dismounting (goal: finish baling as soon as possible)		X		X		X
	Leave tractor and PTO running before dismounting (goal: finish baling as soon as possible)	X		X		X	

A Management-Supported, Worker-Driven System

In the BBS process, workers play critical roles. The two most critical are that they: (a) play a primary role in identifying and describing the list of critical behaviors associated with their work, and (b) serve as observers to record at-risk and safe behaviors and provide feedback to workers and to the organization. Management has to function in a way that encourages and supports the roles that workers play in the BBS system. This is how a supportive safety culture is created, supported and maintained. Each of these major areas is discussed below.

Identifying and Describing Critical Behaviors

The issue of critical behaviors overlaps with the management supported worker-driven safety system concept because identifying safe behavior and correcting at-risk behavior is central to the BBS process. Identifying critical behaviors is sometimes referred to as "targeting behaviors" (Sulzer-Azaroff & Austin, 2000) and "pinpointing" (McSween, 1995). Identifying the safe way to do a job (safe behaviors) is recommended because it facilitates positive feedback from observers and management to workers. Critical safe behaviors are usually assembled into a checklist format. Such checklists can be generic or job-specific (Geller, 1996). A generic checklist is used to observe behaviors that occur during several jobs, such as lifting items, cleaning up work areas, or operating a general-use tool or machine. A job-specific checklist is designed for one job, such as hitching a tractor to a specific machine. In either case, only critical behaviors key to completing the job or task safely should be listed. Both types of lists can be organized or grouped into categories. Precisely defined behavioral definitions must be developed for all checklist items, and are often included on the reverse side of the checklists. Once properly trained to use the checklists, observations generally go smoothly (Reber & Wallin, 1983; Sulzer-Azaroff & Fellner, 1984).

A key point in developing observational definitions, checklists and categories (hereafter referred to as checklists) is that workers develop them, using their own language. This helps to personalize and validate this aspect of the BBS process with workers. In practice, there can be a considerable range of detail among checklists: it is more important that the lists are seen as highly relevant to workers, rather than a more idealized version as might be developed by a safety professional. A second key point is no overlapping of categories. A checklist with overlapping categories might create inconsistency among different observers as they make behavioral observations. Because

BBS is designed as a continuous safety performance improvement plan, it is expected that the development of checklists would be gradually improved upon over a course of a few years. An example of a generic and a job-specific critical behavior checklist are provided in Tables 5.5 and 5.6, respectively.

Critical Behavior Observation and Feedback

The idea of systematically observing workers for critical safe and at-risk behaviors is fundamental to the BBS process. Research suggests outside observers, supervisors, safety personnel and plant workers can all be effective observers and feedback providers, and combinations of these groups are often used (Sulzer-Azaroff, 1982; Krause et al., 1999). One advantage of workers observing coworkers is that it is a good way to introduce trust and confidence in the BBS system. The purpose of ongoing behavioral observation and feedback is continuous improvement in safety performance. Workers are normally more receptive to feedback from peers than from supervisors or safety professionals. Workers often associate observation of safety-related behaviors by management personnel and outsiders with critical, disciplinary and punitive actions, based on past experiences.

The proper use of the observation and feedback technique requires training. This is normally provided by a consultant who specializes in BBS, or by management personnel who themselves have been trained by a consultant. Characteristics of good observers include: high credibility with peers, knowledge about the work to be observed, and good verbal and interpersonal skills. Observation and feedback is often done in teams of two to increase reliability of observations and over time, should involve as many different workers as possible.

Table 5.5. Generic checklist of critical behaviors for machine shop work.

Procedures/Activity	Safe Observation	At-Risk Observation
Wears eye protection		
Uses hearing protection		
Wears foot protection		
Hand tools put away		
Walking aisles cleared		
Bench-working space uncluttered		
Has lifting aids available		
Has lockout/tag-out equipment		
Floor work area free of tripping hazards		

Table 5.6. Specific checklist of critical behaviors for hitching a tractor and wagon.

Procedures/Activity	Safe Observation	At-Risk Observation
Hand signals were used		
Helper kept hands from pinch points during tractor movement		
Safety pin used		
Helper moved from between tractor and wagon before tractor movement		

Not all jobs or tasks will have critical behaviors associated with them from start to end. For instance, when handling pesticides, the operation may involve a combination of pouring a restricted use pesticide liquid into a large tank where it is mixed and transferred to spray tanks for future handling via an enclosed pumping system. The primary exposure to the pesticide for the worker is during the pouring of the liquid into the large tanks. This means that critical behavior would need to be observed only for the pouring of the pesticide into the large tank, rather than, for instance, transporting the pesticide to the point of transfer, or staying around to watch the pesticide mix being transferred to spray tanks.

Feedback on safety performance should be given on the spot, but information developed from the observations can also be used for safety meetings and as public information. Providing feedback means talking to the observed worker about what the observer(s) has just seen and noted on the critical behaviors checklist. As explained by Krause (1997):

> The technique for providing this feedback follows a proven sequence. Positive feedback is given first. The observer talks with the employees about the safe things he saw, emphasizing especially those things that demonstrate improvement over previous observations. The observer then talks about areas that need improvement. Their manner is helpful throughout, making suggestions, asking questions, encouraging questions…, and actively engaging in problem solving with the workers observed.

Loafmann (1997) notes that feedback is just information, or has a neutral value, unless it is used to actually discuss the work observed. By discussing the behavior observations daily or weekly to analyze progress toward ongoing safe behaviors and obstacles to that progress, workers receive reinforcement of safe behaviors, which lead to safety habits that last.

The importance of having developed good critical behavior descriptions and checklists cannot be overstated. A checklist that correctly contains and describes the critical behaviors of a job or task is the substance of the feedback. It provides the observer with specific facts about the task about which to talk to the observed worker. The checklist facilitates a discussion that is objective and based on common ground because the workers developed the checklist. Injuries often occur when workers are doing their routine jobs in a less than safe manner; often unaware of their at-risk behavior because the job is performed almost habitually. Behavioral observation feedback ensures that workers receive information about their safety-related performance. Since the feedback emphasizes positive performance, observer feedback becomes a soon-certain-positive consequence for safe behavior.

A single observation of critical behavior is a sample of workplace safety. It is the accumulation of these samples over time, collected systematically and scientifically, that provides a measure of workplace safety. Trends, or pictures, can be developed for how a site or workplace is changing, either growing safer or becoming less safe.

Besides providing a basis for on-the-spot feedback, observational data can be used to guide discussion at safety meetings, particularly tailgate or other job-site training sessions, for incentive or recognition programs, and to target jobs or tasks that may be in need of revision. If observational data is used in recognition or award programs, the percent of safe behaviors over time should be the basis for the recognition. A job or task that seemingly results in a relatively low percent of safe behaviors suggests a possible system or process flaw that requires revision.

Management Support

The BBS system cannot function unless management creates and supports a climate and culture conducive to a continuous improvement safety performance process. While workers drive the system, it is management that ultimately determines whether a BBS process succeeds or fails because management controls the resources and sets long-term organizational goals. Therefore, training management personnel in BBS methods is as crucial as training the workers. There can be as many different ways of fostering a supportive climate for the BBS process as there are employers and business operations. When done correctly, BBS becomes as integral to the everyday operation of the company as does production and product quality.

Nearly every textbook and major review article on the BBS process lists different key managerial steps or factors for successful implementation. For example, each of the four textbooks previously mentioned organize materials in reference to management strategies and actions differently, as do Hidley (1998), Blair (1999), and Sulzer-Azaroff and Austin (2000) in journal articles. This is partly because BBS is still evolving as a major industrial safety strategy. Hidley's (1998) listing and discussion of seven factors considered critical to the success of BBS processes is typical. These factors and a brief explanation of them are summarized and outlined in Table 5.7.

Successful implementation and maintenance of a BBS process is often marked by three key factors. The first is taking advantage of every opportunity to involve workers in each phase of the process. For example, workers can be a part of each of the seven steps listed above. Recent research suggests that making participation in a BBS process mandatory resulted in higher levels of employee involvement *and satisfaction* with the process (DePasquale and Geller, 1999). A second key factor is that management must follow through with system changes once they have been identified. System changes normally mean that an engineering- or environmental-oriented change is needed in a facility, on a worksite, or perhaps with a job or task itself. For example, to maintain safe housekeeping practices in a farm shop, adequate storage space is necessary to store tools and supplies out of working and walking spaces, step stools to reach materials on shelves, U.L.-approved cabinets for storage of flammable liquids, etc. If a company does not follow through with such changes, it will undermine the entire BBS approach. However, to the extent that such changes are made, they help foster a positive safety culture—the third key factor.

Table 5.7. Seven factors considered critical to successful behavior-based safety processes.

Factor	Description and Meaning
1	Use a process blueprint—a structured, rigorous implementation. • Know the necessary steps; execute them meticulously. • Creatively apply scientific principles; avoid blindly trying "this and that." • Thorough planning and self-discipline are crucial throughout the process. • The path forward has specific steps and a pre-defined sequence of events. • Remain focused on change rather than on attitude-adjustment or consciousness raising.
2	Emphasize communication and buy-in. • Refers to "marketing" within the organization. Includes what, how and when aspects. • Requires executives, managers and labor to jointly develop commitment to a continuous improvement safety performance plan. • Develop key questions to obtain feedback from all levels on the process.
3	Demonstrate leadership. • An organization expresses what it values through actions, not its rhetoric. Key leaders cannot assign "oversight" roles; they must provide it. • Leaders set examples by talking and acting in different ways; they enable new ways to become institutionalized and a part of strategic corporate goals; they have a developmental orientation as opposed to sitting back, waiting to see what employees devise. • Leaders provide resources, benchmarking and teams to solve specific problems, and feedback, evaluation and accountability mechanisms. • Leaders from management and labor do not use safety as a bargaining chip in labor negotiations.
4	Ensure implementation team competence. • Refers to identification and training of steering committee and its leader. • Members of steering committee are "believers" in the BBS system; their role is to identify *for management* existing barriers to continuous safety performance improvement. • The BBS process places workers in new roles; intensive training may be required to match roles with emerging responsibilities and tasks.
5	Use action-oriented training. • Management and workers are trained for their roles and responsibilities immediately before they are to be performed. This facilitates transfer of training and feedback. • Skills acquisition is the focus of training. Practice of skills in a nonthreatening environment is critical. • Avoid overemphasizing technical tools at the expense of interpersonal skills.
6	Use data for continuous improvement. • A successful BBS effort continually measures processes and results. • Measurement should be considered an evolving process because a successful BBS itself should be continually evolving as a continuous improvement process. • Personnel must be skilled in analysis, problem solving, planning and communication. • Systematically removing barriers is where the rubber meets the road in BBS.
7	Provide appropriate technical resources. • Technical resources for site-specific problem solving must be available.

A positive, supportive safety culture flows from successful implementation of a BBS process when the BBS system becomes integrated into the culture of the organization itself. A sign that BBS is becoming, or has become, a part of the organizational culture is when critical behavior problem solving becomes a team process and easily follows a systematic procedure for resolution. Krause (1997) suggests these steps might include:

1. Identifying a problem;
2. Identifying root causes;
3. Generating potential actions;
4. Evaluating possible actions;
5. Developing an action plan; and
6. Follow up, measure and evaluate.

Hidley (1998) summarizes how a successful BBS process results in a positive and supportive safety culture:

> When behavior-based safety works, it can contribute greatly to a firm's well-being across several dimensions. For example, safety becomes more integrated with production and quality. Workforce morale improves as employees see the fruition of their efforts. Incident frequency rate declines and employee involvement in safety increases. Furthermore, workers' compensation expenditures begin a continuous decline as incidents become both less frequent and less severe (p. 30).

Readers are encouraged to consult any of the textbooks or journal articles cited in this section to learn additional details about the BBS process.

Criticisms of Behavioral Models

Scholarly debates among the proponents of the two foundational approaches to human behavior are deeply philosophical, touching often on theological and philosophical histories and basic premises, purposes and methods of science and scientific inquiry. Delving into these issues goes well beyond the scope of this review. And these issues are well covered in many texts, including those by Proctor and Weeks (1990), Barone et al. (1997), Sternberg (1999), Uttal (2000) and Lee (1998). Even so, a brief look at theological and philosophical histories provide an appreciation of these issues.

Advantages of cognitive approaches to understanding human behavior are often entwined in contemporary criticisms of the behavioral approach. In a similar vein, limitations of the cognitive approach are often entwined in contemporary discussions of the advantages of the behavioral approach. Behaviorism as a concept for the study of human behavior crystallized as a specific method during the 1920s-1940s, while cognitive psychology crystallized as a specific method during the 1950s-1960s. Because cognitive psychology arose primarily to address the shortcomings of behaviorism, its very existence and prominence is considered *prima facie* evidence of its advantages. Hence the literature on advantages and limitations of the two foundational approaches is easier understood by looking at what the critics of each approach have said. A brief summary of such criticisms is presented below. The summaries are followed by a discussion of how farm safety and health education may be impacted by limitations and shortcomings of behavioral models.

Cognitivism

The books by Lee (1998) and Uttal (2000) are typical of the criticisms of the literal view of cognitive psychology in general, and the chapter by Norman and Conner (1996) in *Predicting Health Behavior* (Conner and Norman, eds., 1996) and an article by Weinstein (1993) in the journal *Health Psychology* are typical of criticisms specifically applicable to health behavior. Specific criticisms are summarized in Table 5.8, but a few additional comments can be made. Objections to cognitivism are on both philosophical (or theoretical) and methodological grounds. The most basic assumption of the literal view of cognitive psychology is that cognitions of which the person is aware are the immediate causes of human behavior (Lee, 1998). Or to say it more directly, cognitions cause behavior. And since cognitions occur before behavior, an assessment of relevant cognitions can be used as the basis for predicting behavior. It follows that if we can predict behavior, then we can also change or modify undesirable behavior toward more desirable behavior (who decides desirability is set aside for the moment). Implicit in the assumption that behavior can be predicted and changed is that the behavior is explained and understood.

A literal view cognitive approach suggests the real world is relevant only peripherally as the raw material from which the individual constructs a subjective world (Bruner, 1986). Therefore efforts to understand social behavior should focus on cognitive interpretations of events and not the events themselves. Uttal's (2000) objection to literal view cognitivism is that cognitions are essentially inaccessible. In his words, "Can the mind be observed, measured, and then analyzed into its parts?...is it accessible to scientific examination?" (p. 1). This is a philosophical question for which there may never be a completely satisfactory answer. But if, as many behaviorists insist, the mind is impossible to study scientifically because it is unobservable (inaccessible), then it is also true that theories about how the mind works—in other words cognitive theories—can never be completely proved or disproved because the theories themselves are hypothetical constructs (Lee, 1998; Uttal, 2000; Weinstein, 1993). Saying this differently, the mind does not exist in a material or physical sense for the world to see, nor can an individual see his or her own mind. Therefore, others or our self can only infer or hypothesize whatever is in the mind.

Such views are at the heart of a philosophical debate between behaviorists and literal view cognitivists about what makes for good behavioral science. The hallmark of behaviorism is its insistence on empirical observations under controlled experimental laboratory conditions as the standard of scientific inquiry (O'Donohue & Kitchener, 1999). Elliott Aronson (1999), a social psychologist, recently supported this notion about scientific inquiry when he wrote that the "first step in the scientific process is observation…the next step is to make a guess as to why that happens" (p. 429). What this approach to scientific inquiry is suggesting is that the interest of science is best served when the phenomenon of interest are observed first, and then explained through hypotheses development and experimental testing. Literal view cognitivism, as explained above, develops hypotheses and experiments and then makes observations to confirm or deny the hypotheses. While this is a practical approach for theory development, it does not necessarily follow good science (McQueen, 1996). McQueen noted that theories of human behavior, despite showing weak relationships between psychological concepts such as motivation, beliefs, attitudes, opinions and actual behavior, seldom start with behavior.

A second and closely related philosophical point of debate is the issue of analyzability (reducibility) (Uttal, 2000). Two assumptions of all literal view cognitive models are that whatever is in the mind can be reduced into component parts for study, and that interrelationships between and among component parts can be understood (Lee, 1998; Uttal, 2000). For example, when the concept of self-efficacy was proposed by Bandura (1977), he essentially reduced the variable "expectation" into "efficacy expectations" and "outcome expectations," and then further divided "efficacy expectations" into "magnitude," "generality" and "strength" components (constructs). Later, Meier (1983) proposed a "reinforcement expectation" construct, which he suggested should come from the "outcome expectations" construct, in effect, reducing that construct.

The propensity of literal view cognitive scientists to keep reducing what are essentially hypothetical constructs into additional hypothetical constructs is one basis for the proliferation of cognitive theories and models (Weinstein, 1993). Lee (1998) argued, "proposing the existence of more hypothetical variables misses the basic problem that the variables are not adequately defined, their relations are not defined, and we have no satisfactory way of assessing them" (p. 12). Cognitive behavioral models are operationalized primarily through survey research and self-reports. There is a considerable body of research suggesting that a participant may misunderstand questions, interpret the same question differently from the researcher and from other participants, have false memories about events, respond to the same question differently depending on presentation variables, suffer from a variety of biases (e.g., self-evaluation, attribution errors, probability biases, etc.), and be unaware of his or her reasons for routine thinking and behavior while responding to surveys about beliefs, thoughts, motivations and behaviors (Garnham & Oakhill, 1994; Nisbitt & Wilson, 1977; Nisbitt & Ross, 1980; Schober, 1999; Sedikides, 1999; Shum & Rips, 1999; Sirken & Schechter, 1999; Schwarz, 1999; Tourangeau, 1999).

Table 5.8 summarizes limitations and shortcomings of common literal view cognitive theories of human and health behavior. The questions asked by Weinstein are taken directly from his article (Table 2), while the Norman and Conner comments represent the authors' summation of Chapter 7 in their book, *Predicting Health Behavior*. Most of the criticisms emanate from methodological problems inherent in attempts to explain and control the inner workings of the human mind. There are two purposes for this table. The first is to reinforce the idea that, despite their popularity, there is still considerable debate about the usefulness and accuracy such models purport to tell us about human thinking and subsequent behavior. Secondly, the table will help inform the discussion of how problems with cognitive theories may impact farm safety and health education.

Behaviorism

Critics of behaviorism as a theory for explaining and controlling human behavior have been long-standing and numerous (Lee, 1998). For example, Mahoney (1974) argued that the view that only the physical world exists was overly simplistic. Dissatisfaction with Skinner's radical behaviorism (and with Watson's behaviorism before that) helped to spur the approach to behavioral study that became known as cognitive psychology (Proctor & Weeks, 1990). Cognitive psychology, or cognitivism, is defined by Sternberg (1999) as, "…the belief that much of human behavior can be understood if we understand first how people think" (p. 75). This definition embodies the most fundamental difference between the two foundational blocks of psychology: behaviorism posits that human behavior can be explained and controlled without detailed explanations of the workings of the mind; mentalism posits that human behavior cannot be satisfactorily explained or controlled without understanding the human mind.

It is easier to summarize the criticisms of radical behaviorism (or of Skinner's operant conditioning) than cognitivism because only one model of operant conditioning (Antecedent-Behavior-Consequence) is essentially relevant to this review. This is not saying that there are not other models of contemporary behaviorism. For example, "Teleological Behaviorism," "Biological Behaviorism," "Logical Behaviorism," "Rye's Behaviorism" and others are discussed in *Handbook of Behaviorism* (O'Donohue & Kitchener, eds., 1999). Operant conditioning behaviorism, though, is the form used to date in connection with occupational safety and health.

A common theme that runs through many objections to behaviorism is the "anti-human" approach to human behavior. Behaviorism views humans as having few essential differences from laboratory animals. Cognitivists of the constructivist view strongly reject this (Lee, 1998). The basis of cognitivism (human uniqueness) can be traced back to the time of Greek physician and philosopher Hippocrates. Hippocrates believed the mind was a distinct entity that resided in the brain and it—the mind, not the brain—controlled the body (Sternberg, 1999). In the "mind-body dualism" view, the body was composed of physical substance, but the mind was not, and the agent of

Table 5.8. Limitations and shortcomings of cognitive behavioral models.

Source	Limitations and Shortcomings
Weinstein, 1993	1. To what extent do beliefs about the likelihood and severity of health outcomes combine to influence behavior? Are their effects additive, multiplicative, or some other function? 2. How important are perceptions of risk compared to beliefs about other nonrisk issues in influencing behavior? Are there categories of nonrisk issues (such as emotional experience, self-esteem, social approval, etc.,) that are useful to distinguish? Are there certain nonrisk issues that are always relevant, or is each health threat unique? 3. Do doubts about one's ability to carry out a precaution have a different impact on behavior than do other obstacles (such as perceived costs and inconvenience)? Does the expected costs of carrying out a precaution need to be considered separately from the expected loss of internal and external rewards resulting from abandoning current behavior, or can they be combined? 4. Do people derive notions of risk-reduction effectiveness from separate beliefs about the risks presented from current behavior and from the recommended alternative, or do they have notions of effectiveness that arise separately? 5. What variables intervene between intentions and actions? Is reaching a decision to take an action a discrete event that can be defined by a dichotomy, or should decisions to act be described along a continuum? Are there other distinct stages to the precaution-adoption process that can be identified? 6. Do dimensions of risk other than likelihood and severity (e.g., familiar vs. novel, visible vs. invisible, imminent vs. delayed) need to be considered? How do they affect behavior? 7. What attributes of precautions other than the beliefs about cost and effectiveness considered in these models (e.g., past patterns of behavior, complexity of action, need for repletion, elimination of risk vs. reduction of risk, etc.,) influence behavior?
Norman & Conner, 1996	1. Only a small amount of variance is actually accounted for in most predictive theories and models of human behavior. 2. Some models lack specificity between cognitive variables and predicted behavior (TRA/TPB are notable exceptions). 3. There is often a large time gap between the completion of questionnaires and measurement of target behavior, leading to contamination by intervening variables. 4. Frequently, behavior is measured on a single occasion in a single context, meaning that the measured behavior may be idiosyncratic or atypical. 5. Many studies have different operational definitions of a theory's major constructs, leading to different measurement scales and little attention paid to reliability and validity of the measures. 6. Belief statements collected from subjects may be biased because subjects are often supplied an initial list of beliefs to select from rather than generating their own list. The supplied list itself may be unrepresentative. 7. Statistical analyses are usually based on "between-subjects" comparisons. This informs us about who may change behavior, but does not inform us about how individuals come to their decisions or about alternative courses of action. 8. There are many research findings that suggest variables not generally accounted for in present theories predict behavior, for example, self-predictions, moral norms, anticipated regret, self-identify, and habit (past behavior). 9. Most health behavior models assume a rational, deliberative processing of information. There is considerable research evidence that this assumption does not adequately reflect intentions, decisions-to-act, or behaviors in many cases. 10. Most health behavior models do not consider that different cognitions may be important at different stages of health behavior change. 11. Many people who intend to perform a behavior fail to do so. Most models do not directly address the issue of failing to translate intentions into action.

control was within the body. Greek philosopher Plato (ca. 428-348) furthered this view of humans by postulating that reality resided not in concrete objects, but in abstract forms of the objects as we thought of them in our minds. For Plato, because true reality lay in abstract forms, pursuit of truth was best found by using philosophical reasoning and speculation analysis.

Sternberg (1999) traces this line of reasoning from Hippocrates and Plato through the beginning of Christian religious beliefs (<A.D. 450), the Middle Ages (A.D. 400-1300), the Renaissance (1300s-1600s), the Nascence (Birth) of Science (ca. 1400) and the beginnings of the modern period (ca. 1600). The idea of "humanism," or humans "as the measure of all things" evolved and solidified. Trains of thought that developed and survived through these periods of time included: (a) the mind and body are separate entities; (b) some knowledge was innate to humans; and (c) that the introspective, reflective method was superior to empirical methods for finding truth. Philosophers and theorists, like Aristotle (384-322), Francis Bacon (1561-1626) and John Locke (1632-1704), disagreed with this view of humans and science. Their view was that reality: (a) is a unified whole, existing in a single plane, rather than in a dual plane; and (b) lies only in observable, concrete objects that the body can sense. From these beliefs flowed the ideas that the study of the mind and the study of the body are one and the same thing, and that humans are born without knowledge and therefore seek knowledge through empirical observation.

These philosophical and theological traditions still mark contemporary criticisms of behaviorism. Hickey (1994) believes that hostility to behaviorism arises because behaviorism runs counter to a philosophy that human action is determined by the soul, spirit or mind, while Hineline (1994) believes behaviorism challenges the notions of free will among human beings. By reviewing the items in Table 5.9, readers can obtain a relative sense of how strictly cognitively oriented psychologists commonly view behaviorism today. The sources of the lists in Table 5.9 are Skinner (1974) and Kimble (1996). Skinner himself has provided one of the most comprehensive lists of common criticisms against behaviorism in his book, *About Behavior ism* (1974). A comparison of the two lists show that the nature of the criticisms has not changed appreciably over time. For a more detailed explanation of these criticisms, and rebuttals to them, consult the texts by Skinner and

Table 5.9. Common criticisms of behaviorism.

Skinner, 1974, pp. 4-5.	Kimble, 1996, p. xi.
1. It ignores consciousness, feelings and states of mind.	1. Behaviorism sacrifices mind, purpose, thought and human experience at the alter of stimuli and responses. It rejects everything that is mentalistic, thus everything that psychology is supposed to be about.
2. It neglects innate endowment and argues that all behavior is acquired during the lifetime of the individual.	2. Behaviorism's stimulus-response approach is atomistic. By its very nature, behaviorism cannot deal with complete individuals or total situations.
3. It formulates behavior simply as a set of responses to stimuli, thus representing a person as an automation, robot, puppet or machine.	3. Behaviorist explanations are reductionistic: Human behavior is either materialistic biology or abstract mathematical equations.
4. It does not attempt to account for cognitive processes.	4. The laws that behaviorism seeks are mechanistic laws of passive adaption. There is no place for human beings conceived as self-directed coping, causal agents.
5. It has no place for intention or purpose.	5. The behaviorists approach is nomophetic. It deals with averages and promotes the concept of the "standard man." It neglects the variance in behavior and fails to recognize that every human being is a unique individual.
6. It cannot explain creative achievements—in art, for example, or in music, literature, science or mathematics.	6. Behaviorism is "scientistic," not scientific. It presumes to measure human attributes that are not quantitative. Its laboratory methods are artificial. They dissect behavior from its natural context and yield results that have no useful applications.
7. It assigns no role to self or sense of self.	7. Behaviorism is simplistic. It lacks the complexity required to capture the subtle nuances and richness of its subject matter.
8. It is necessarily superficial and cannot deal with the depths of the mind or personality.	8. Behaviorism's contributions to understanding of the human condition are trivial. They are a catalog of small effects produced by insignificant causes.
9. It limits itself to the prediction and control of behavior and misses the essential nature or being of man.	9. Behaviorism turns human beings into lower animals. It is insensitive to the scope of human potential and blind to the essential human quality in all of us.
10. It works with animals, particularly with white rats, but not with people, and its picture of human behavior is therefore confined to those features that human beings share with animals.	10. Behaviorism is without human values, without a conscious, without morality or ethics.
11. Its achievements under laboratory control cannot be duplicated in daily life, and what it has to say about human behavior in the world at large is therefore unsupported meta-science.	
12. It is oversimplified and naïve, and its facts are either trivial or already well known.	
13. It is scientistic rather than scientific. It merely emulates the sciences.	
14. Its technological achievements could have come about through the use of common sense.	
15. If its contentions are valid, they must apply to the behavioral scientist himself, and what he says is therefore only what he has been conditioned to say and cannot be true.	
16. It dehumanizes man; it is reductionistic and destroys man *qua* man.	
17. It is concerned only with general principles and therefore neglects the uniqueness of the individual.	
18. It is necessarily antidemocratic because the relation between experimenter and subject is manipulative, and its results can therefore be used by dictators but not by men of good will.	
19. It regards abstract ideas such as morality or justice as fictions.	
20. It is indifferent to the warmth and richness of human life, and it is incompatible with the creation and enjoyment of art, music and literature, and with man's love for one's fellow men.	

Kimble or see Uttal (2000), who comments on these and others' criticisms of behaviorism.

Behavioral Models and Farm Safety and Health Education

A question asked earlier, "Under what circumstances might cognitive and behavioral theories and models of human behavior be applicable to farm safety and health education?" can now be addressed. It should be obvious from the previous sections that there are many constraints associated with theories and models that attempt to explain human thinking and behavior, and some theories may be seriously undermined by the realities of the *farm safety—risk paradox*. For example, the popular cognitive theories and models used with safety and health behavior, though operationalized in real world settings, have almost always been developed from controlled laboratory settings involving hypothetical behavior and outcomes, nonhazardous environments, and college students (Ajzen & Madden, 1986), so relevance in the heat of the moment with real physical trauma or death a heart beat away is always an issue. Acknowledging difficulties with theory and theory development is not to suggest that theory, particularly for farm safety and health education or intervention, is not valued or to be pursued. Following that train of thought is to surrender to the lowest common denominator of agricultural safety and health—every person for himself or herself. This is, to state the obvious, unacceptable. It may be useful at this juncture to remember why theory and theory development in health behavior is important. McQueen (1996) reminds us:

> First, theory provides the guiding light and tells us in broad terms what to study (and what not to study).... Second, adopting a theoretical approach helps define the appropriate methods...if we wish to study the dynamics of behavior we need...methodology which collects behavioral data.... Third, theory serves the useful function of defining that which we cannot understand...thus...we recognize that there are simply some 'black boxes' which are inexplicable at present. Fourth, a theoretical perspective helps us understand how different pieces fit together and why we should value some combinations of variables more than others...(p. 32).

The foundational approaches to the study and control of human behavior have been explained and criticized from conceptual and methodological perspectives. Below, each theoretical approach is considered more practically from the perspective of farm safety and health education

Cognitive Theory

Identified earlier were models of literal view cognitive theory that have been applied to farm safety and health problems and issues. For example, Gjerde et al., (1991) applied the Health Belief Model (HBM) to improve the health knowledge, attitudes and self-reported behaviors of swine confinement workers; Lee (1995) used the Theory of Planned Behavior (TPB) to understand factors that influence parents' decisions to expose children to major farm hazards; and Witte et al. (1993) used the Persuasive Health Message Framework (PHMF) for a formative evaluation of farmers' safety practices and beliefs about farm equipment accidents and safety. These and earlier citations suggest important data regarding farm safety and health beliefs and practices can be collected on a variety of specific issues. Of agricultural studies found using a specific cognitive behavioral model, only Gjerde et al. (1991) and Petrea (1996) have attempted to experimentally evaluate a behavioral outcome using a specific behavioral model (the HBM and TPB, respectively). Both studies included self-reports of behavior change by pork producers and found increased use of respiratory protection after an educational intervention. Overall, the scarcity of cognitive theory-driven studies in agriculture precludes in-depth analyses of cognitive-behavioral models based on actual experience. Even so, some practicality issues, which are closely related to methodological issues, can be raised to encourage discussion and debate among agricultural safety and health professionals.

While it seems innately normal to want to understand why farmers think and behave as they do, remember that the ultimate goal of safety and health education is behavior *change*. To borrow from injury epidemiology (Haddon, 1970), it may be that an in-depth *understanding* of why farmers think as they do is not necessary to effect important behavior change. Furthermore, it must be remembered that even though a specific human behavior may be the last discernable action before an injury incident, it does not automatically follow that human behavior is the most logical or efficient point of attack for injury prevention (Haddon, 1970; Robertson, 1983). Therefore, behavioral models that purport to *explain* human thinking and behavior might seem to stand in direct opposition to injury epidemiology. But this isn't true because injury epidemiology recognizes the need for a holistic approach to injury prevention, including safety education (Christoffel & Gallagher, 1999). Clearly, some context is needed and it will be offered shortly.

Literal view cognitive theory by nature is individualistic, that is, data is collected from individuals in the attempt to understand how individuals think and behave (Sternberg, 1999). Theories such as TRA and TPB proposition that, "any single sample of behavior reflects not only the influence of a relevant general disposition, but also the influence of various other factors unique to the particular occasion, situation and action being observed. By aggregating different behaviors, observed on different occasions and in different situations, these other sources of influence tend to cancel each other, with the result that the aggregate represents a more valid measure of the underlying behavioral disposition than any single behavior" (Ajzen 1991, p. 180). People tend to think and behave in a generally consistent way across similar situations, and it is this consistency that can be used to understand and predict behavioral intentions. The same data is collected for a group of people, for example, farmers or perhaps just dairy farmers, and pooled to make behavioral predictions for the entire group. As noted earlier, the theories built on this proposition have enjoyed considerable research support. But what does this mean in the context of farm safety and health education?

An individual's "general pattern of thinking and behavioral *intentions* across similar situations" may have little relevance to actual behavior in the face of farm hazards and everyday work. This may be why Aherin (1987), in his TRA study, found neutral or negative mean values for four of six behavioral intentions while their corresponding attitude and subjective norm mean values were significantly more positive; Petrea (1996) was able to explain only 28 percent of the variance in his TPB study; Lee, while accounting for two-thirds to three-fourths of the variance in her TPB study (Lee et al., 1997), did not find support for the construct of perceived behavioral control (Lee, 1995); Witte et al. (1993), in their PHMF study that incorporated TPB constructs, found farmers consistently reporting that they use safety measures *in general*, but did not use specific safety measures for some important sources of fatal injury; and Marlenga and Lee (1996), in their HBM study, reported that among farm men and women who had received a presumptive skin cancer diagnosis, few acknowledged practicing sun-safe behaviors four to six months later.

One important disconnect between health behavior theory and actual farm work behavior is that many health behavior theories are predicated on the assumption that people anticipate a negative health outcome and the desire to avoid the outcome or reduce its impact motivates self-protective behavior. The lack of injury in most cases of at-risk behavior is at odds with the first portion of the assumption, and what farmers say in focus group interviews is at odds with the second part of the assumption. The lack of relevance between general patterns of behavioral intentions and actual behavior may stem from complex interactions involving time pressures, economic issues, old machinery, fatigue, weather, risk perceptions, injury experience, social or group norms, and the like. It is not likely that such issues are satisfactorily accounted for in the standardized questionnaires used to assess behavioral intention, which the farmer may complete while sitting at the kitchen table drinking an iced tea, or in the standardized coding systems for analyzing their responses (Mishler, 1986). This scenario represents a combined practical/methodological problem not easily overcome.

A second practical/methodological issue is the resource costs of implementing behavioral models research. As any student—or even casual reader—of behavioral models quickly comes to understand, correctly implementing such models is a time- and labor-intensive process laced with opportunities to commit methodological errors. Additionally, the nature of expectancy-value theories like TRA and TPB is that it is practical to collect data on only a few behavioral intentions at a time. The extensive variability of agricultural operations and work makes this an important practicality issue. While acknowledging any type of survey research can be time and labor-intensive, and considering the limitations of such theories in explaining intended and actual behavior (see Tables 5.1, 5.8, 5.9 and related discussions), the question arises, "Is there a more efficient way to get to the same or similar information?" There may be.

Several studies were identified in the Nature of Farming section of Chapter 4 (see the two paragraphs following Table 4.1) that discussed the *farm safety—risk paradox*. These studies and reports, generally more limited in purpose (e.g., there is no formal prediction of behavioral intention or behavior) provide considerable insight into the beliefs, feelings and thoughts about work hazards and related behavior (Lee et al., 1997 is the only study cited in this section and the Nature of Farming section). These studies used mail surveys (e.g., Elkind, 1993; Whitman & Field, 1995), telephone interviews (e.g., Schwab et al., 1995; Hwang et al., 2000), and focus groups (e.g., Green, 1999; Kidd et al., 1996) to collect data. Qualitative analysis methods (e.g. focus groups, structured interviews, key informants, etc.,) are particularly suited for assessing culturally sensitive beliefs, values and knowledge data (Fetterman, 1989; Morse & Field, 1995), and for formative evaluations and program design (Dean, 1994; Rossi et al., 1999). In addition to Kidd et al. (1996) and Green (1999), Rhodes and Hupcey (2000) and Thu et al. (1998) have recently used qualitative assessment strategies to collect farm safety and health beliefs data.

While all research methods have limitations, focus groups in particular have advantages with farm audiences. They can be very time and cost efficient, type III errors (external validity) are minimized, and farmers have a rich oral tradition with regard to farm work (Kidd et al., 1996). A type III error concerns inappropriate conclusions being drawn because a program was not implemented as intended, or there were considerable differences in how a single program was implemented at different sites (Basch et al., 1985; Scheirer, 1994). These errors are minimized in focus groups because of constant comparisons of statements and reverification of meanings with the participants as the focus group meeting progresses (Lincoln & Guba, 1986). Focus groups and other qualitative assessment methods are not burdened with making formal statistical connections between beliefs, attitudes and behavior. Focus group methods allow researchers more flexibility in exploring important cultural and situational issues that more fully explain underlying beliefs, values, feelings, attitudes, etc., that are important to the issue being studied. Unlike with TRA/TPB, behavior data can be collected in the same setting as other data.

It is not suggested that cognitive behavioral and health behavior models be abandoned in the effort to better understand and direct farm safety and health education, nor that focus groups or similar qualitative research methods be used exclusively. But it is suggested, for conceptual, methodological and practical reasons, concerning literal view behavioral models: (a) there are important limitations in their explanatory power for connections between farm safety beliefs, attitudes, behavioral intentions and behavior; (b) most models predict behavioral intention, whereas the phenomenon of most importance is actual behavior; and (c) there are less resource intensive alternatives for collecting very similar data. As a result, it may be that such theories are best utilized for the most important, far-reaching and complex educational issues facing agriculture. Examples might include tractors and ROPS, children and farm safety, cultural farm safety and health practices, or farm safety and health public policy changes. Using intervention evaluation models such as the Intervention Effectiveness Research (IER) framework described in Chapter 4 can help researchers and educators decide when behavioral theory and models would be efficacious and efficient.

Behavioral Theory

From a farm safety and health education perspective, behavioral theory is more tightly connected to a specific behavioral model, Behavior-Based Safety or BBS, than are cognitive models. Hence practical considerations between farm safety and health and BBS can be more directly discussed. Before doing this, it is noted that the BBS process is not without its critics. Eckenfelder (1997), Smith (1999) and Ragan (1997) are among those who have written critical articles on BBS. Eckenfelder and Smith are private safety consultants and Ragan is a corporate safety director. Eckenfelder believes that working directly with corporate culture (values and attitudes) is a more effective approach to a lasting improvement in safe work behavior. Smith is critical of the science of behaviorism in general, and advocates a quality management systems approach to safety. Ragan, too, believes that a whole-systems perspective is a more appropriate approach to safety than is a focus on BBS. Similar themes run through all three articles, and closely mirror general criticisms of behaviorism as a method of behavioral change.

No studies were found where a BBS process had been used in a production agriculture setting. So the question of its effectiveness in production agriculture must be speculative at this point in time. Even so there appear to be few inherent structural impediments to a production agriculture operation implementing a BBS process. In an article on myths, magic and reality of BBS, Blair (1999) points out the importance of customized implementation of the BBS processes, and that you do not need a consultant to implement BBS. On the other hand, nearly every proponent of

BBS points out how it is neither a "quick fix" nor an approach to injury prevention that can be taken lightly if it is to succeed. Given the culture of risk taking prevalent in agriculture, and the many factors that contribute to the *farm safety—risk paradox*, it might seem imprudent to suggest that BBS could work in agriculture. But a closer look at operational principles of BBS from the perspective of agricultural settings helps to clarify this statement. The discussion will focus first on operational principles as applied to small and/or family farm operations, and then to larger farm operations.

Small and/or Family Farm Operations

At its core, and from the perspective of production agriculture operations, the BBS process means:

1. Identifying and operationally defining critical safety-related behaviors,
2. Observing the behaviors to gather data on the frequency of those behaviors,
3. Providing feedback, and
4. Using the gathered data for continuous improvement (Krause, 1997).

All of these steps are well within the capabilities of most adult farm operators, workers and families. Even when a farmer is working alone, a self-observation technique can be used (Krause, 1997; McSween, 1995), negating the need for a separate feedback step. Remember that the BBS process, as practiced in industry, incorporates several aspects of organizational behavioral management. To the extent that farm operations are small and/or strictly family operated, most organizational issues become irrelevant, or at a minimum, considerably simpler. For example, looking again at Table 5.7, it would be inherently less complex to operationalize those seven factors with two or three workers than with 50 or 100 or more workers.

This does not mean that an appropriate, or quality, implementation of the seven factors, or of the other core steps, is any less important when applied to small or family-operated businesses. In one sense, the lack of an organizational hierarchy means fewer people can contribute to the development, implementation and maintenance phases of the BBS process. Fewer people mean more responsibility on less shoulders. Understanding the principles and methods of identifying and defining critical behaviors, behavioral observation and feedback, and using the data to drive continuous improvement in safety performance, requires training—perhaps more training than most farm operators would feel they have time to devote to farm safety. And farm operators with few labor resources may feel too constrained by daily and seasonal work routines to devote time to critical behavior identification and observation. There may also be resource limitations for system safety improvements. Finally, stable and nonchanging work environments in other industries contribute to BBS successes whereas small and/or strictly farm family operations are often marked by constantly changing work environments. Issues such as time, training, resources and constantly changing work environments may indeed limit application of a BBS process on real farms, particularly small and/or family-operated farms. But reasons such as these are already present among agricultural operators and operations, and are not unique to the BBS concept.

It is not farfetched to suggest that the major impediments to small and/or family farm operations implementing a customized (to their operation) BBS process will emanate more from a lack of perceived need for such a process, and a lack of comfort with such a formalized approach to farm safety. The absence of a serious injury in the past is likely to contribute to this perspective. But more deeply seated feelings may underlie the reluctance to try a BBS process. For instance, a general resistance to any kind of major operational change, uncomfortable with observing and commenting on the safety behavior of others, or a fear that the procedure could lead to confrontation with another worker, could all be reasons why a farm operator would resist seriously considering a customized BBS process. Father-child, sibling, or child-grandparent relationship impediments are even more likely in family-only operations. Such motivational barriers would indeed be difficult—if not impossible—to overcome. Still, small and/or farm family operations could benefit from a BBS process because it provides a system that would: (a) identify physical and behavioral hazards; (b) remind all workers how to perform work safely; (c) provide a mechanism for safety training; (d) improve personal relationship skills; and (e) in the case of a hired employee or volunteer laborer, provide protection against a personal injury lawsuit.

Large Farm Operations

Large farm operations, defined as those with $100,000 of farm products sold in a year, are becoming more common. For example, in 1978 farms with $100,000 sales were 9.8 percent of all farms; by 1997 they comprised 18.1 percent of all farms (USDA, 2001). Larger farms use more hired and contract labor than smaller farms, which rely heavily upon self and unpaid family labor. For example, farms with $500,000 or more in sales and 1,000 acres or more accounted for approximately 70% and 48%, respectively, of total labor expenditures (Runyan, 2000).

As production agriculture establishments increase in size, they increasingly take on characteristics of other industrial work sites. Examples of these characteristics include, but are not limited to:

- More paid, organized and full-time labor;
- Work environments that are more standardized;
- Different personnel for management and labor functions;
- More levels of management (e.g., executives, site managers, foremen);
- More operations with vertically integrated products;
- Increased likelihood of falling under the jurisdiction of state and federal safety and health regulations;
- More safety and health support from the insurance industry; and
- Increased exposure to legal liability for preventable injury.

Behavior-based safety processes have been used successfully in industrial settings with small groups of workers (<35) including with a sugar cane machinery manufacturer (Reber et al., 1984) and a shipyard (Saari & Nasanen, 1989). To the extent that a production agriculture operation relies on full-time hired labor and has standardized work environments, full-time managers or supervisors, and a named safety and health department or staff, the more similar it will be to industrial operations where a BBS process has proven successful.

Production agriculture operations that rely on migrant labor present different, though not insurmountable, challenges. Language, worker turnover, and a piece-rate pay structure (e.g., lbs. or bu. harvested, no. of plants planted) are obvious difficulties. Still, the specific jobs and tasks performed by migrant labor do not change dramatically from one season to another, and reports such as *Simple Solutions: Ergonomics for Farm Workers* (Baron et al., 2001) illustrate that tasks can be redesigned to improve safety while maintaining productivity. Several work sites have specifically addressed ergonomic issues through a BBS process (Geller, 1996; Krause, 1997).

Chapter Summary

To summarize a few important points of the discussion so far:

- It was mentioned that few modern psychologists ascribe to the extreme version of either foundational approach to behavior change (radical behaviorism or mentalism);
- A multitude of strongly oriented literal view mentalistic theories and models have dominated the psychological literature for the past few decades, and these theories and models have dominated attempts to understand and explain human behavior and behavior change;
- Some of these dominant behavioral theories and models are the ones that have been applied to agricultural safety and health issues in the past decade;
- The relevance of the principles of the behavior-based safety process in structured and controlled workplaces cannot be dismissed out of hand; and
- There is considerable dissatisfaction among psychologists of all stripes with many of the dominant cognitive and behavioral theories and models of behavior change.

It is important to recognize that the concepts, theories and practices of cognitive psychology, its many subdisciplines, and behaviorism are greatly intertwined. For example, O'Donohue and Kitchener (1999), in the preface of *Handbook of Behaviorism*, noted that Edward Tolman (1886-1959), who saw himself as a staunch behaviorist (in terms of methodology), is now considered a member of the cognitivist family by many researchers. O'Donohue and Kitchener went on to write:

> Where should we put Tolman? Arguably, he was as much of a behaviorist as a cognitivist. But if so, then how do we distinguish behaviorism from cognitivism? …it is our view that no one has yet set out the features distinguishing behaviorism from cognitivism (p. xx).

Another example is seen in Geller's (1996) *The Psychology of Safety*. Geller's "actively caring" concept (purposeful and directed behavior for the benefit of others) is an integral part of his applied behavioral analysis approach to improving safety performance. Geller, a strong BBS advocate, supports development of his actively caring concept with several constructs that are central to cognitive psychology, such as self-esteem, self-efficacy and empowerment. And Barone et al. (1997) in their book, *Social Cognitive Psychology, History and Current Domains,* states: "Many constructs in today's social cognitive psychology are derived not from earlier perception and learning schools but from the cognitive revolution in psychology, begun in the 1950s. The information-processing tradition provided the computer model of cognition and a new terminology of inputs, processing, outputs, heuristics, and so on, which is now pervasive in social cognitive psychology" (p. 22).

The next chapter reviews learning theory and applications to farm safety and health education. Importantly, behavioral and learning theories are closely linked throughout their histories, with both dating back to Aristotle and Plato. Merriam and Clark (1993) efficiently convey this in their statement, "It is commonly understood that learning is central to human behavior, and that life itself presents never-ending opportunities for learning" (p. 129). After learning and learning theory are discussed, an alternative approach (the constructivist view) to farm safety and health education is identified. A main feature of the alternative approach is that cultural and social issues relevant to occupational agriculture safety and health are more successfully embedded within the approach, which in turn helps to harmonize behavioral theory with learning theory and farm safety and health education.

Glossary

Associationism. A movement in psychology that examines how events or ideas become associated with one another in the mind to result in a form of learning. Focuses on stimulus and response effects. A forerunner of behaviorism.[g]

Attribution theory. A theory of how people assign causation to events.[a]

Behaviorism. A philosophy of the science of human behavior that focuses on observable and environmental aspects of behavior. Is considered an extreme version of associationism. A meta-philosophy that contains lower-level theory such as learning theory and evolutionary theory.[f]

Behavior. A specific action that occurs in a particular setting at a particular time.[e]

Cognitive psychology. The study of the mind and how thinking and reasoning influences behavior.[c]

Confirmatory bias. The tendency to attend to, gather and remember information in a way that supports preconceptions, stereotypes and initial impressions.[a]

Constructivist tradition. A line a development leading to today's social cognitive psychology that emphasizes cognitive interpretation via schemas and their ongoing transformation by experience.[a]

Culture. A relatively organized system of shared meanings.[e]

Expectancy-value theory. A theory that predicts behavior from expectancies (beliefs) and values (goals, needs) rather than associations and reinforcement.[a]

Experimental psychology. Preceded cognitive psychology. Goal was to make the old associationists-empiricists philosophy of mind into an empirical and experimental science.[d]

Functionalism. The first U.S.-born movement in psychology. Developed from structuralism. Addressed the broad questions of how and why the mind works as it does by seeking functional relationships between specific earlier stimulus events and specific subsequent response behaviors.[g]

Gestalt psychology. The word gestalt refers to a unified whole; a pattern or organized field having properties that cannot be derived from the summation of component parts. Gestalt psychology refers to the doctrine that psychological phenomenon do not occur through the summation of individual elements, but through gestalts.[d]

Goal. A cognitive representation of a desired future outcome, product or end state.[a]

Goal theory. A theory of self-regulation concerned primarily with explaining and predicting the effects of various types of goals and goals-relevant feedback on performance.[a]

Hindsight bias. Tendency to overestimate powers of prediction once a person knows the outcome of a given event.[b]

Information processing tradition. A line of development leading to today's social cognitive psychology that uses computer programs to simulate cognitive processes.[a]

Introspection. Preceded behaviorism. Observation of one's own mind at work; a carrying out of a mental task while trying to observe the mental processes that underlie its performance.[d]

Locus of control. Construct for attributions about the source of reinforcement, either internal or external.[a]

Mentalism. A philosophy of the science of human behavior that focuses on mental or cognitive aspects of behavior. A major focus is on thinking and reasoning. A metaphilosophy that contains lower-level theoretical constructs such as perception, motivation, attitudes and intentions.

Motive. An affectively charged "push" to work toward and derive satisfaction from a certain broad class of goals.[a]

Operant behavior. A unit of behavior for which no correlated stimulus can be detected *a priori*, but which can be correlated with immediate environmental consequences.[f]

Operant conditioning. Shaping or directing behavior through the use of reinforcers. Reinforcers are any consequential events that modify the likelihood that the operant behavior will occur again.[f]

Overconfidence effect. Confidence in one's judgment that exceeds its accuracy, typically resulting from several bias and errors in judgment, leading one to believe, mistakenly, that one is an accurate perceiver and predictor of other people and events.[a]

Pragmatism. An outgrowth of functionalism. Pragmatists are concerned not only with knowing what and why people do what they do (functionalism), but also with what they can do with the knowledge of that information (e.g., teach, predict behavior).[g]

Schemas/Schemata. Symbolic constructions representing events, guiding psychological functioning, and changing in response to experience.[a] Internal representations of knowledge; the way people organize knowledge of the world.[h]

Self-efficacy theory. A theory of self-regulation concerned primarily with understanding the influence of beliefs about personal mastery or self-efficacy in self-regulation and goal attainment.[a]

Self-regulation. The automatic and deliberative processes by which people control and direct their cognitions, emotions and behaviors in pursuit of their goals.[a]

Self-theory. A set of images and propositions about oneself.[a]

Social cognitive psychology. The study of cognitive and social processes as dual aspects of human transactions.[a]

Social constructive-development theory. Adults (experts) guide children (novices) in the co-construction of meaning.[a]

Social gestalt tradition. A line a development leading to today's social cognitive psychology which extended Gestalt perceptual theory to the study of cognitive and personal processes in a social context.[a]

Social learning tradition. A line of development leading to today's social cognitive psychology which joined animal learning theory with clinical formulations about personality and psychotherapy.[a]

Social psychology. The influences that people have upon the beliefs or behavior of others.[b]

Structuralism. The first major school of thought in psychology. Developed in Germany. The goal of structuralism was to understand the structure of the mind by analyzing it (the mind) into its constituent components or contents.[g]

Values. Universal statements about what we think is desirable or attractive. Values do not ordinarily contain statements about how they are to be realized.[e]

Sources: [a]Barone et al., 1997; [b]Aaronson, 1999; [c]Neisser, 1967; [d]Garnham & Oakhill, 1994; [e]Smith & Bond, 1999; [f]O'Donohue & Kitchener, 1999; [g]Sternberg, 1999; [h]Halpern, 1996.

6 Learning Theories, Models, and Applications

This chapter explores learning theories and models, and their application to agricultural safety and health. Like behavioral theory in general, learning theory can be approached from a cognitive or a behavioral perspective, and similar to behavioral theory in general, learning theory derives a majority of its contemporary theories, models, principles and concepts from behaviorism, and more recently, cognitivism. Literature shows that learning theory quotes many of the same founding fathers, for example J.B. Watson, E.L. Thorndike, J. Dewey, B.F. Skinner and J. Piaget (e.g., Burns, 1995; Sutherland, 1997), and use many of the same underlying concepts, such as goal theory, motivation, attention and self-efficacy (e.g., Silverman & Casazza, 2000; Sinnott, 1994, Steers & Porter, 1991) as general behavioral theory. It was pointed out in the previous chapter that many principles and constructs of behaviorism and mentalism overlap and entwine.

Davis et al. (2000) argue that, from the perspective of contemporary learning theory, both behaviorism and mentalism make the same limiting assumptions: (a) thought occurs deep inside the individual's body; (b) behavior and cognition are mechanical processes, thus they follow rules; and (c) full understanding of a learning event can be obtained by reducing it to its most basic components. Kirshner and Whitson (1997) have stated, "Cognitivism, like behaviorism, understands knowledge and learning as resulting from experience within a stable, objective world. Community and culture can enter into cognitivist theory only insofar as they are decomposed into discrete elements…" (p. vii). Kirshner and Whitson also stated that such a focus was severely limiting and noted dissatisfaction with it among educators, anthropologists, psychologists and social theorists.

Greeno (1993) has noted that "educational practice in the United States is shaped fundamentally by the view of cognition and learning that…focuses on individual knowers and learners who acquire knowledge and cognitive skill by adding small pieces incrementally to what they have learned previously…" (1993, p. 154). This approach to learning has been termed the "empty vessel" or "banking concept of education," where teachers deposit education into the minds of students (Freire, 1971). The banking deposit approach implies several things about the educational process, including: (a) a teacher teaches and the students are taught, hence knowledge becomes an object to be bestowed on students by teachers; (b) concepts such as "truth" and "reality" become what the teacher perceives it to be and teaches, rather than what may actually be; and (c) an educator's role is to organize the way the world (i.e., reality, truth, knowledge) enters into the student. While this concept of education is equally descriptive of both formal and informal education, informal education as it applies to agricultural work safety and health is the primary focus of this chapter.

A consideration of learning theory and informal education from the perspective of agricultural work safety and health brings into focus four issues that can help to clarify important facets of learning theory and farm safety and health education. These are:

- Terms and definitions,
- Learning from experience,
- Adult learning theory, and
- Applicability of traditional learning theory to farm safety and health education.

After these issues are discussed, a consideration of an alternative approach to learning and farm safety and health education will be identified.

Terms and Definitions

A number of terms from the fields of learning theory and occupational safety and health are used interchangeably or within a synonymous context, but that nevertheless may mean different things to readers. The terms education, training, learning and knowledge, among others, are used in both fields but do not have sharp definitional distinctions. An understanding of how these terms are used in relation to learning theory is important for understanding new approaches to farm safety and health education identified at the conclusion of this chapter. Few authors give precise, dictionary-type definitions of such terms, choosing rather to explain them by examples or to assume that they are satisfactorily defined by the context in which they are used.

For example, Miller, in his book *Objective-Based Safety Training*, states, "…I will use the words *education* and *training* interchangeably…since I believe that training is a basis for education and cannot be separated" (1998, preface). Miller devotes two chapters (Chapters 5 and 6) to learning, including adult learning theory. Hilyer et al. (2000), in their book *Effective Safety and Health Training*, never actually define the words education, training or learning, but specifically discuss learning objectives and by implication, learning as it applies to behavioral models. On the other hand, NIOSH (1999a) states, "*Training* is widely understood as communication directed at a defined population for purpose of developing skills, modifying behavior and increasing competence. Generally, training focuses exclusively on *what* needs to be known.... *Education* is a longer-term process that incorporates the goals of training and explains *why* certain information must be known" (p. 5). NIOSH does not specifically define learning, but views it as a dependant variable in their Training Intervention Effectiveness Research (TIER) model (p. 7).

Davis et al. (2000) note the many different ways learning is thought of in the context of learning theories: "...learning as *acquiring and inputting data*, to learning difficulties as problems in *brain wiring*, to communication as *transmission and interfacing*, to thinking as *processing and compiling information*, to memorizing as *storing*, and to remembering as *retrieving and outputting*" (p. 53). May (1998) noted the distinction Canadian adult education educators made between learning and education. She summed up their perspectives by noting the simplistic distinction that tends to equate education with formal schooling and learning with both formal and informal schooling. May quoted Jarvis (1995), who had previously made this distinction when he stated, "Learning is something which people do, while education is a social institution which provides learning opportunities for people" (p. 9). Tight (1996), in his review of key concepts in adult education and training, notes "...there is no common understanding of how we learn, and there are many and divergent opinions amongst psychologists and educators about just what learning is" (p. 21).

The term knowledge is similarly abstract. Davis et al. (2000) noted that knowledge is commonly viewed as object-based. Knowledge is something to be *acquired, grasped, held, possessed, exchanged* and *wielded.* This metaphoric framing suggests knowledge can be located either outside (*objective* knowledge) or inside (*subjective* knowledge) the knower, and in this way, learning "becomes a process of bridge-building—a matter of connecting the external to the internal" (p. 53). Tight (1996) noted that the term knowledge is more associated with the broader term education, while the same substance or information is more likely thought of as a skill if it is used in the context of training. Hettich (1997) traces different forms of knowledge in his review of cognitive development among college students. Among others, he cites Baxter Magolda's (1992) work as establishing a framework for knowing that included four levels of knowledge:

1. Absolute (knowledge is certain),
2. Transitional (some knowledge is uncertain),
3. Independent (knowledge is mostly uncertain), and
4. Contextual (knowledge is uncertain).

Bereiter (1994) distinguished between learning activity and *knowledge building*, in the sense that the purpose of formal schooling is not improving the minds of students, but improving the knowledge that is being collectively created within their minds. Clearly, educators and learning theorists have different perspectives on knowledge.

Candy (1991) contrasted knowledge from the traditional education perspective with the constructivist view. The traditional educator view is that knowledge is an accumulated body of empirically verified facts, derived directly by observation and experimentation. This view leads to the assumption that there is an objective reality to which learners should be introduced. The constructivist view is that people construct knowledge as they interact with the world, and this construction forms the building blocks of knowledge and understanding (Spigner-Littles & Anderson, 1999). Several implications are associated with this view. First, knowledge is largely shaped by individual perceptions and personal experiences. Second, learning by a person often proceeds from a series of personal propositions that, if not disproved, are assimilated into explanatory schemas "as though" they were demonstrably true, and once thoroughly internalized, for all intents and purposes, they are true for the individual. Third, knowledge is largely a socially based process because people live and interact in a social world.

Fourth, knowledge is both public and private, and exists in three domains: (1) the physical and social environment governed by observable rules (one form of public knowledge), (2) the cultural environment governed by norms negotiated by individuals living in the culture (a second form of public), and (3) emancipatory knowledge (private knowledge). Emancipatory knowledge is personal and involves self-awareness and how one sees himself or herself in various life roles and social expectations. Fifth, considering that much knowledge is experiential (interacting with the world), social and public, learning becomes an "active process of constructing a system of meanings and then using these meanings to construe or interpret events, ideas or circumstances" (Candy, 1991, p. 278).

The concept of experience is closely connected to knowledge. Experience in any endeavor provides knowledge about that endeavor. From an agricultural safety and health perspective, most of us have wondered why teenagers, parents, operators and other adult farm workers don't seem to learn effectively—from the professional's viewpoint—about safe and at-risk behaviors and practices based on real experiences with farm work hazards, risks and injury, whether their own or those of family, neighbors or friends. A review of the connections between experience, knowledge and learning helps to clarify this particular issue.

Learning from Experience

The idea of life experience in learning theory is very much like that of knowledge: on the surface everyone understands it, but beneath the surface, there are numerous ways to conceptualize, define and explain it. The perspective that is important here may be summed up in the question, "Why do we learn more from some life experiences than others?" Or stated another way, "Why do some life experiences have a greater impact on us than others?" The current literature on this topic often begins with a discussion of the philosophy and writings of philosopher and learning theorist John Dewey (1859-1952) (e.g., Jarvis, 1987a; Merriam, 1994; Starnes, 1999). Among other things, Dewey pointed out that experience included an active and a passive element, and their peculiar combination. The active element refers to "trying" while the passive element refers to "undergoing." It is in the connection between the two that constitutes their peculiar combination and what characterizes learning from experience. In the words of Dewey (1924):

> When we experience something we act upon it...then we suffer or undergo the consequences. We do something to the thing and then it does something to us in return; such is the peculiar combination.... The connection of these two phases of experience measures the...value of the experience. Mere activity does not constitute experience.... Experience as trying involves change, but change is meaningless transition unless it is consciously connected with the return wave of consequences which flow from it. When an activity is continued *into* the undergoing of consequences, when change made by action is reflected back into a change in us, the mere flux is loaded with significance. We learn something.... It is not experience when a child merely sticks his finger into a flame; it is experience when the movement is connected with the pain which he undergoes in consequence...sticking the finger into flame means a burn. Being burned is a mere physical change, like the burning of a stick of wood, if it is not perceived as a consequence of some other action... 'to learn from experience' is to make a backward and forward connection between what we do to things and what we enjoy or suffer from things in consequence (p. 163-164).

In today's literature this learning from experience is referred to as gaining tacit or practical knowledge. Tacit knowledge is practical know-how not usually taught directly or openly expressed. It is the kind of knowledge one acquires on a job or in everyday kinds of situations (Torff & Sternberg, 1998). For example, a young teenager on a farm may figure out for himself or herself the best method to hitch a tractor and wagon when there is no one around to help. Practical knowledge has three characteristic features (Sternberg et al., 1995). First, it is procedural in nature, taking the form of "knowing how" rather than "knowing what." Second, it has practical usefulness because it is normally directed toward goals that people value. And third, it is normally acquired under conditions of low support. That is, it is gained on one's own with little or no instruction. Gaining practical knowledge is at the center of learning from experience but does not explain how learning from experience occurs.

Kolb (1984) developed the concept of "experiential learning," defining learning as, "the process whereby knowledge is created through the transformation of experience" (p. 38). Kolb developed a four-stage cyclical model of learning from experience that further explains how people make connections and learn from experiences. The first stage is the experiencing of the event, happening or occasion. This is followed by reflection upon the event (stage two), which involves thoughts and feelings. The third stage involves developing tentative concepts and generalizations that explain or accommodate the experience to one's self. The fourth stage is then testing the tentative explanations and generalizations in new situations. The new situations, of course, again represent stage one, hence the reference to cyclical stages.

Jarvis (1987a), building upon the work of Dewy and Kolb, has defined learning as, "the transformation of experience into knowledge, skills and attitudes" (p. 164). Jarvis (1987b) also suggested nine possible responses to a life experience: three that are nonlearning, three that represent nonreflective learning, and three that represent reflective learning. The three nonlearning responses are termed: presumption (unconscious behaviors, little if any conscious thought); nonconsideration (the experience not paid attention to due to lack of time, interest, distractions, etc.); and rejection (person deliberately chooses to reject learning from the experience). The nonreflective learning categories are: preconscious (incidental learning, learned once and then hardly thought about again); practice (development of physical and social skills); and memorization (storing of knowledge for future use). The three categories of reflective learning are: contemplation (intellectual reflection); practice reflection (reflection in action, practical knowledge); and experimental (connection between knowledge and reality is demonstrated).

Merriam (1994) has noted the theories and writings of Dewey, Kolb and Jarvis agree that all life experiences offer the potential for learning, but that the ingredient of attention and reflection must be present for people to transform experiences into learning. The idea of attention and reflection, in turn, provide a link between experience and meaning-making. On the one hand, if an experience is too familiar with previous experiences, a person pays it little attention or ignores it completely (e.g., a minor slip on a manure-slick barn floor that has no serious consequence). However, if an experience is too incongruent or different from all previous experiences, a person may reject it out of fear or discomfort and not learn from it (e.g., a first tractor overturn incident where the operator is not injured). Therefore an experience must contain the right amount of novelty, difference, discomfort or puzzlement for it to be "meaningful" (Jarvis, 1987a). The degree to which a given experience is novel or puzzling can vary significantly among individuals and for each individual at different times in his or her life. It follows that experiences contribute to learning only when an individual considers them meaningful to himself or herself.

Much of the literature on connections between meaningful life experiences and learning can be summarized by the series of statements below. These connections between life experiences, meaning-making, and learning are logically viewed as circular rather than linear (Davis et al., 2000). They all feed into each other rather than the connections proceeding from one to another along a straight line. These statements have been developed primarily from the Jarvis (1987a) and Merriam (1994) references.

- Experience relates to the process of living in society, and involves a relationship between people and the sociocultural milieu in which they live. Therefore learning is also related to that sociocultural milieu.
- Experiences occur in sociocultural-temporal situations, that is, within a specific context. No experience has meaning in itself: it must be given meaning by an individual. Hence experiences are "interpreted" from within the context in which they occur.
- In everyday experience, individuals may be barely conscious of routine experiences as they rely upon a biologically-based stock of knowledge from previous experiences. As long as there is continuity between people's individual stocks of knowledge and the sociocultural-temporal world in which they act, they are able to perform in an almost automatic fashion. In this way, experiences act as a reenforcement to the stock of knowledge already acquired, and does not add to the stock of knowledge.
- Experiences that reenforce stocks of knowledge but do not add to it, increases one's confidence to perform similar actions in the future.
- Life experiences can be viewed as life events that are normative or nonnormative. Normative events lend predictability to life. A nonnormative event is disruptive but can create an opportunity for learning from the event.
- Learning from experience is about transforming experiences into knowledge.
- Transforming experience into knowledge can occur when an experience is critically reflected upon. Critical reflection draws upon both the current experience and previous experiences, and includes a consideration of feelings, values, ideas and behaviors associated with those experiences.
- Critical reflection by an individual occurs primarily when an experience personally affects the individual and is subjectively valued by him or her. It follows that what may be a meaningful experience to one individual may not be to another.
- One outcome of critical reflection is whether the person views the experience as normative or nonnormative.

Practical experience is a key concept in adult education theory (Burns, 1995; Long, 1998). It is one of the central ideas that separates child learning (pedagogy) from adult learning (andragogy). The notion that teaching and learning are fundamentally different enterprises for children and adults is most commonly attributed to Malcolm Knowles and his books *The Modern Practice of Adult Education: From Pedagogy To Andragogy* and *The Adult Learner: A Neglected Species* (1970, 1973). Although this view has been challenged (e.g., Hanson, 1996; Long, 1998; Norman, 1999; Tight, 1996), and is barely, if at all, mentioned in more contemporary approaches to learning (e.g., Davis et al., 2000; Dominice, 2000; Gee, 1997; Silverman & Casazza, 2000), it is common to see the concepts of andragogy referenced in textbooks on adult learning and training in workplaces (e.g., Burns, 1995; Hilyer et al., 2000; Miller, 1998). Therefore, it needs to be briefly discussed here in the context of traditional and contemporary approaches to adult learning theory. This will be followed by discussion on the applicability of learning theory to farm safety and health education.

Adult Learning Theory

Learning about occupational safety and health implies a focus on adults because adults have occupations. The most relevant adult learning theory literature is found under subject headings such as adult education, adult learners, lifespan learning, learning and development, and other similar titles. As mentioned, the literature that broadly reviews approaches to adult learning theory generally breaks along the lines of behaviorism and mentalism, though authors have different ways of grouping and naming mentalist approaches. For example, Burns (1995) groups according to the behaviorist (stimulus-response), the cognitive-Gestalt, and the humanist/phenomenological approaches. Miller (1998) uses the categories of behaviorist, cognitivists, humanists, and social learning. Scott (1998), on the other hand, lists liberal/perennialism, progressivism, behaviorism, humanism, and radical or critical orientations. The specific differences among the mentalist groupings are less important than understanding that most mentalist learning theory mirrors mentalist behavioral theory.

The behaviorist views learning as synonymous with behavior change. In short, if behavior has not changed, learning has not taken place. The physical and social environment determines learning, and the role of the teacher is to arrange and control the environment so that learning takes place. The three primary forms of learning are incidental (also called observational, habituation), classical conditioning, and operant conditioning (also called instrumental), with operant conditioning seen as the most pertinent to adult learning (Geller, 1996; Sutherland, 1997). Behaviorism and operant conditioning were explained in detail in Chapter 5. Educational terms and strategies that are represented by this approach include competency-based education, mastery learning, and assertiveness training. Importantly, objections to behaviorism as learning theory are one and the same with objections to it as general behavioral theory (e.g. Burns, 1995).

As suggested by the various groupings of mentalist approaches, the mentalist approach to learning theory has evolved a much wider range of concepts, principles and constructs than the behaviorist approach. Malcolm Knowles, with his Theory of Andragogy, was the first to articulate a specific theory of adult learning. Andragogy (the art and science of teaching adults) was distinguished from pedagogy (the art and science of teaching children) by the five assumptions that underlie Knowles theory (summarized from *The Adult Learner: A Neglecected Species*): four were in the first edition (1973), with a fifth added in the third edition (1984). These were:

1. As a person grows and matures, his (or her) self-concept moves from one of total dependency to one of increasing self-directedness.
2. As an individual matures, he (or she) accumulates an expanding reservoir of experience that causes him (or her) to become an increasingly rich resource for learning, and at the same time provides him (or her) with a broadening base to which to relate new learning.
3. As an individual matures, his (or her) readiness to learn is decreasingly the product of his biological development and academic pressure, and is increasingly the product of developmental tasks required for the performance of evolving social roles.
4. Children have been conditioned to have a subject-centered orientation to most learning, whereas adults tend to have problem-centered orientation to learning.
5. Internal factors rather than external factors are the primary motivators for learning.

Knowles (1980) also suggested that teaching adults has seven stages. Taken from the second edition of *The Modern Practice of Adult Education: From Pedagogy To Andragogy*, these are:

1. Setting a climate for learning, which includes both physical and psychological factors.
2. Establishing a structure for mutual planning.
3. Diagnosing learning needs.
4. Formulating directions (objectives) for learning.
5. Designing a pattern of learning experiences, which includes continuity, sequence and integration.
6. Managing the learning experiences.
7. Evaluating and rediagnosing learning needs.

Similar to general behavioral theory, theories of learning have proliferated in the past two decades, with one theory arising from the bits and pieces of another. For just one example, Boud et al.'s (1996) "model of reflection in the learning process" relies heavily on Mezirow's (1991) Theory of Transformative Learning, which draws from, among others, Kelley's (1963) Theory of Personal Constructs, Habermas's (1971) knowledge framework, and Roth's (1990) notion of perceptual filters (Boud et al., 1996; Cranton, 1998). The number of different theories and models and their intricacies are beyond the scope of this review, but the review by Silverman and Casazza (2000) in their text *Learning and Development* is helpful in illustrating the breadth and connectivity of modern learning theories.

Silverman and Casazza (2000) consider learning and development as it applies to any type of postsecondary education (e.g., college level, formal, informal, workplace). Their approach to learning is that: (a) no one viewpoint or explanation of behavior can fully account for learning, (b) learning occurs in a variety of settings and learning in one setting often affects the learning that occurs in another setting, (c) learning is best experienced in settings that acknowledge the uniqueness of individuals, and (d) using a variety of instructional approaches enhances learning. From their examination of learning theory research, Silverman and Casazza organized the research into six theoretical foundation topics that identify theories and concepts by: (a) personal development and learning, and (b) cognitive development and learning. Silverman and Casazza acknowledge that their six topics were not the only topics related to learning, but that these were primary as they related to the practice of learning and development.

The six topics are shown in Table 6.1, along with the theoretical constructs the researchers considered to relate to each topic. The purpose of Table 6.1 is not only to show the range of constructs and topics in adult learning theory, but also to illustrate the interplay between learning theory and general behavioral theory. Readers will recognize several of the theories and concepts as having been discussed in Chapter 5, and may recognize several others as topics discussed in the general literature of learning and teaching.

Table 6.1. Foundational learning theory constructs by research topics.[a]

Foundational Topic	*Theories and Concepts*
Personal Development and Learning	
Self and identity	Cultural interdependence Autonomy-interdependence Identity development stages Meaning systems and critical reflection
Motivation	Locus of control Attribution Self-worth Self-efficacy Self-concept Hierarchy of needs
Interaction with environment	Behavior-person-environment Student-environment reciprocity Model of social climate Theory of margin Learning styles Environmental coping strategies
Cognitive Development and Learning	
Ways of knowing	Cognitive apprenticeship Tacit knowledge Multiple intelligence Emotional intelligence Triarchic theory of intelligence Zone of proximal development Constructivism Field dependence-interdependence Models of reading Schemata Cognitive development
Learning styles and preferences	Cultural suicide Collaborative learning Communication and culture Cultural identity Culturally responsive teaching Left brain-right brain research Learning style inventory Personality-based indicator
Self-regulation and goal setting	Metacognition Comprehensive monitoring Self-regulated learning Goal orientation

[a] Adapted from Table 1 in Silverman and Casazza, 2000, p. 56.

Applicability Issues

It was noted above that few contemporary learning theorists make a sharp distinction between adult and child learning theories (e.g., Hanson, 1996; Davis et al., 2000; Silverman & Casazza, 2000). Learning theory literature, however, adult and otherwise, approaches learning from the perspective that a teacher (educator, instructor, facilitator) is interacting with a learner (student) on some regular frequency, whether it be in a formal academic setting (e.g., a university degree or credit course program), an informal setting (e.g., noncredit course for personal growth), or in an occupational work setting. But neither a focus on adults nor regular learning interactions suitably describes the circumstances or context for most instances of agricultural safety and health education (vocational agriculture classes may be an exception to this statement). For example, a focus on adult education ignores young workers, and even adults with young children who are exposed to the farm work environment, and most instances of teaching agricultural safety and health to adults or children is a one-shot presentation. One-shot presentations that are often of limited duration (e.g., one hour or less) and mixed in with other subject matters (e.g., crop nutrient management, tillage practices, animal production, etc.) seriously undermine applications of learning theory at the time of the presentation. For example, there is little opportunity to take advantage of the different learning styles and preferences that undoubtedly exist in any group. This leads to the question, "How does learning theory apply to the realities of farm safety and health educational practice?" The answer is that traditional learning theory has little relevance to modern farm safety and health education practice. Identifying the nature of the interaction between those doing the education and those being educated help explain this view.

From the perspective of traditional learning theory, the educator is viewed as an "expert" and a learner is viewed as a "novice." Therefore, this relationship can be viewed from a power relation framework (Merriam & Caffarella, 1999). Because the educator has technical or subject matter expertise, and this knowledge is wanted or needed by the learner, the educator has power over the learner and thus controls the agenda of the educational task. Expertise from this framework is in two categories: subject matter and in the methods of training and education. An expertise in one does not necessarily imply expertise in the other (Burns, 1996), although many audiences expect (or perhaps, hope) for it. The power relation framework has historically been—and continues to be—the structure of formal schooling (Candy, 1991). It is no surprise that it was extended to less formal settings and occasions where education was desired or needed: this extension was, after all, the basis for the formation of Cooperative Extension (Seevers et al., 1997). And as agricultural work injuries mounted, it is no surprise that Cooperative Extension, other agricultural organizations, and public health educators applied prevailing learning theory to farm safety and health prevention efforts. In the earliest days of farm safety and health education, for example the 1940s into the 1970s, the power relation framework, with its reliance on experts and expertise, was less problematic than it is today for several reasons.

First, most farm safety and health experts in the earlier days had more natural credibility with farmers than do experts today. This is because the experts that most often interacted with farmers were themselves still active in farming, or had considerable experience with agricultural work growing up on the farm. Specialty crops and large livestock confinement operations were fewer in those times, so there was more commonality of experience among those growing up on farms and those still farming or helping with work on farms. University experts from Cooperative Extension and Farm Bureau safety leaders in particular directly interacted more often with farmers through county and state farm safety committees. There was also much less safety information available in earlier days because the field of farm safety was itself still developing, and formal research into the causes of farm accidents or the effectiveness of injury prevention strategies, programs and projects was less widely understood.

Another crucial factor was that communication and publication capabilities were considerably more limited and constrained, creating an environment where what the expert knew about safety, really was news to the learner. The environment of the early years can be contrasted with today's environment where the experts may have little if any experience or connection to any working farm. Technological revolutions (e.g., 300 hp articulated tractors, computer chips to control individual dairy cow feed makeup and availability, genetically modified seeds, global positioning systems), specialty farms (e.g., raising emus, organic crops, ginseng farms), and contract farms operated by corporate entities (e.g., Tyson Farms; Purdue, Inc.; Hanover Foods, Inc.) make it difficult for farmers themselves to feel connected to each other. And, the World Wide Web, a CD-ROM, or a cellular phone means information on any topic is a few clicks or seconds away.

Finally, as documented in Chapters 2 through 4, contemporary farm safety and health educators have had very little that can be considered new to say about agricultural work safety and health since at least the 1970s. Tractors, machinery, animals, falls, grain dusts, agricultural pesticides and other chemicals, children and aged workers, migrant and seasonal laborers, exclusion of farming from OSHA regulations, and promotion of voluntary safe behavior have been broadly defining the exchange between farm safety and health educators and learners for a minimum of 25 years (the time since the OSHA small farm restriction), and actually much longer than this for most of this list. New or innovative ways to convey the same essential information (e.g., farm safety day camps, machinery simulators and interactive CD-ROMS to teach about machinery safety), or a few new pieces of equipment arriving on the scene (e.g., ATVs, high-speed tractors), doesn't negate the fact that to many farm workers, the essence of the information presented (e.g., be careful, shut off the equipment before dismounting, one seat-one rider) is all too familiar.

Adding to the issue of expertise and credibility are the relative infrequency of contacts with learners, the absence of methods expertise—few farm safety and health educators can be considered professional trainers—the infrequency of actual injury from at-risk behavior and therefore the consistent reinforcement of at-risk behavior, and the incredibly difficult job of overcoming the *farm safety—risk paradox*, it is no wonder that traditional learning theory seems limited for farm safety and health education.

An Alternate Framework

Narrative psychology provides educators with a fundamentally different approach to farm safety and health education. It is consistent with contemporary learning theory in its learner-centered approach; it provides a methodology for taking advantage of the expertise that resides in each farm worker; it does not exclude valuable contributions to understanding and learning provided by traditional behaviorism and cognitive psychology; and it more readily acknowledges the roles that sociocultural influences have on individuals to conform to cultural norms.

Before proceeding further, a few words about overall context are important. Recall that education (or persuasion) is one of three broad lines of attack for injury prevention and control (Christoffel & Gallagher, 1999). Modifying the environment (engineering, engineering technology) and legal requirements or prohibitions are the other two broad strategies. As a rule, education is not considered the most effective or efficient approach because it relies on individual behavior each time a hazard or risk is encountered. Engineering and legislative remedies are considered more effective and efficient methods because they can be more universally and permanently applied to eliminate or control hazards and risk. There is no suggestion in this chapter that these fundamental rules of injury prevention are not applicable to agricultural injury prevention. At the same time, practical and sociopolitical realities, such as the geographic dispersion of work sites and the exclusion of agriculture from most OSHA regulations, suggest that education oriented strategies must be employed to a greater extent than in other high-risk industries.

One reason for this is that with many agricultural applications, engineering-oriented strategies typically require a strong and ongoing educational component. For example, with ROPS for tractors, education is needed to help convince farmers to add them to older tractors, not remove them when they are already on a tractor, and to buckle the seatbelt to receive the most protection from the ROPS. And education is needed to help convince farmers they should not remove guards from machinery drivelines or disable bypass starting safety devices. Additionally, as explained in Chapter 4, safety education has practical value for promoting positive farm safety cultural values and beliefs. To the extent that the educational method is used, it is incumbent upon all agricultural injury prevention professionals to continually seek ways to more effectively utilize it.

Narrative psychology is representative of the cognitive constructivist approach (a similar term is the constructivist tradition in Barone et al., 1997) to understanding and learning (Mayer, 1996; Rossiter, 1999) and represents just one new way of approaching education and learning. A constructivist approach encompasses the idea of situated cognitions (explained below), which is an idea that is not without detractors. For example, Bereiter (1997), in his article critical of situated cognitions, points out how situated cognition has not lead to new learning theory, and refers to a "collaborative knowledge building" approach. Tripp (1996) wrote an article critical of several aspects of situated learning, including the constructivists' view of knowledge and the diminished role of didactic instruction. Still, given the constraints and circumstances of the traditional approach to farm safety and health education, there are several aspects of the cognitive constructivist approach and of narrative psychology that seem particularly relevant to improving the effectiveness of the educational approach for agricultural audiences. With these caveats in mind, the work of Dr. Henry (Hank) Cole surfaces as an attractive alternative.

Cole, an educational psychologist, has conceptualized an approach to farm safety and health education that embraces constructs from behaviorism, cognitive psychology, cultural learning, and learner empowerment theory. Narrative psychology provides the framework for Cole's model (1997). The model, originally labeled, "a cultural, cognitive, and behavioral model for health beliefs and safety behavior" (Cole, 1997), later shortened to the "Theory of Narrative Representation" (Cole, 2000), is too new in its application to farm safety and health education to have a developed a supportive empirical evaluation record. Cole's model, though, comes from his work in mining safety where supportive empirical evidence is found (Cole et. al., 1988; Cole et. al. 1993). His model addresses theoretical objections to the common behavioral models identified in Chapter 5, and also important aspects of the *farm safety—risk paradox*. Cole's model is presented in Figure 6.1. Three constructs—narrative psychology, enculturalization, and empowerment education—important to Cole's (1997) conceptual model have not yet been discussed and are briefly reviewed below. The 1997 article by Cole provides the most complete explanation of the model.

The attractiveness of Cole's (1997) model for farm safety and health education is that it provides a method to capture the reality of at-risk and safe behavior of those we are attempting to help, and combines it with insights provided by traditional behavioral and health behavior change models that may explain limited but important aspects of human thinking and behavior. The Cole model does not attempt to predict future safety and health behavior; rather it helps us understand past and current behavior.

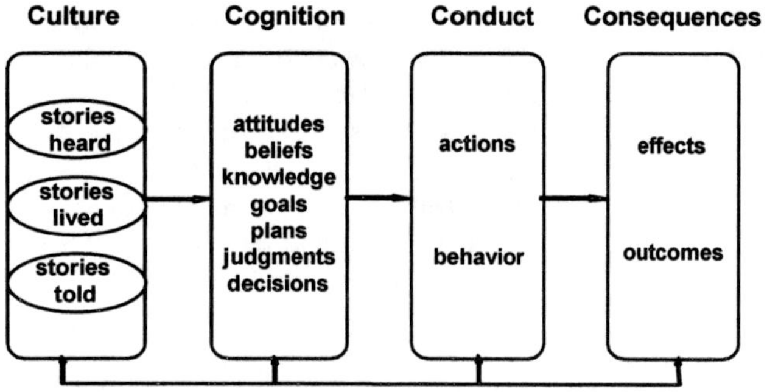

Figure 6.1. A cultural, cognitive, and behavioral model for health beliefs and safety behavior.

Narrative Psychology

Narrative thinking has been proposed as a root metaphor for psychology (Sarbin, 1986). A root metaphor conceptualizes and organizes how people think about, or frame, their understanding of human action. As a comparison, cognitive psychology's root metaphor has been the mind as a computer and thinking as a computational function. This is more commonly referred to as "the human as an information processor" (Bruner, 1990; Mayer, 1996; Sarbin, 1986). Under this metaphor, the human mind has been seen as accepting inputs (taking in information), processing the data (e.g., interpreting, reacting), and producing output (e.g., language, thinking, feelings, behavior). Mayer (1996) has referred to cognitive psychology's information-processing theories as learning theory's second metaphor (the first was learning as a function of strengthening the stimulus-response associations). Mayer (1996) proposes that learning theory has recently developed a third metaphor termed "cognitive constructivist."

Cognitive constructivism grew out of educational psychology's attempts to understand learning and development in real world situations, as opposed to the contrived and artificial laboratory settings of information-processing theories (Neisser, 1967; Mayer, 1996). Cognitive constructivism suggests knowledge is constructed from direct, active experiences in life, and it exists in a person's mind as mental models or pictures (schema, representations) (Bower & Morrow, 1990; Smilkstein, 1990-1991). According to this theory, mental models and representations in the mind grow out of attempts to make meaning of the world as people experience it. It is through these mental models that strategic plans are developed, which guide human actions toward desired goals and end results (Bower & Morrow, 1990; Mayer, 1996). Cognitive constructivism is closely related to two additional concepts and ideas that provide support for narrative as a root metaphor for human thinking, learning and behavior. These are "situated cognitions" and "storytelling." These concepts, as well as the concept of narrative psychology itself, flow from the premise that *culture* is an antecedent to human thought and behavior (Lonner & Adamopoulos, 1997). Culture is defined here as "a relatively organized system of shared meaning" (Smith & Bond, 1999). Cole has referred to many of these concepts and ideas as "socioculturalism" (Cole, 2001).

Situated Cognition

Many methods of didactic education (teacher lecturing to the student; the power relations framework) assume a separation between knowing and doing. This assumption implicitly treats knowledge as an integral, self-sufficient substance, theoretically independent of the situations in which it is learned and used (Brown et al., 1989). The situated cognition view is that the activity or situation in which knowledge or learning *occurs and is used* is not separable. Rather, the activity or situation is an integral part of what is learned. Considerable evidence for this position comes from linguistics research involving how children learn language and begin communicating to others (e.g. see Berman & Slobin, 1994; Bruner, 1990; Kagan 1981; Mancuso, 1986). Miller & Gildea (1987) have explored how children and adults learn words outside of formal classroom settings. They note that children and adults generally learn words in the context of ordinary communication, and in contrast to formal word learning (e.g. dictionaries and example sentences), thousands of words are learned per year versus a few hundred words per year through formal schooling.

Prominent among the intricacies of language are indexical words (they point to a part of the situation in which the word is being used) that depend on the situation they are being used in for meaning (Brown et al., 1989). For example, words like *I*, *here*, *now*, *next*, *tomorrow*, *afterwards* and *this* are always context-dependant. This is why people wait for or read the rest of the words in a sentence before trying to interpret what an indexical word means. In situated cognition, language—and thus knowledge—are inextricably a product of the activity and situations in which they are produced.

Resnick (1987) has contrasted how situated cognition (learning) differs between most formal educational settings (school) and the everyday world. Her comments provide a good explanation of what situated cognition/knowledge/learning means in everyday terms, and is very relevant to farm safety and health education. Resnick identifies four ways that school (formal) learning differs from learning that occurs outside the school system. The first way is a focus on *individual cognition in school versus shared cognition outside* of school. The dominant form of school learning, performance, and homework is individual, and individuals are ultimately judged as passing or failing based on what they can do by themselves. In contrast, most activity outside of school is socially shared. Work, personal life and recreation all take place within social systems, and each person's ability to function successfully depends to some extent on others. This is what is meant by "socially shared knowledge and skill." This concept is further explained by the example of navigational practice in the highly technological work environment of U.S. Navy ships. Piloting a large naval ship out of a busy harbor:

> ...involves six people with three different job descriptions. On deck two people take visual sightings on predetermined landmarks using special telescopic devices mounted on gyrocompasses that yield exact readings of direction. They call out their readings to two other people, who relay them by telephone to a specialist on the bridge. This individual records the bearings in a book and repeats them aloud for verification. Next to a recorder, another individual uses specialized tools to plot the ships position on a navigational chart and to project where the ship will be at the next fix and beyond. These projections of position are used to decide what landmarks should be sighted next by those on deck and when a course correction will be required. The entire cycle is repeated every one to three minutes (p. 13).

The point of this example is that no individual in the system can pilot the ship alone. The knowledge necessary for successful piloting is distributed throughout the navigational system. Additionally, important aspects of that knowledge are built into the tools used by the navigational crew—telescope and gyrocompasses are a part of the knowledge system. Even though the navigational crew did not build the tools, and the builders of the tools did not help to navigate the ship, they are all a part of the knowledge system for successful piloting of the ship.

The idea of knowledge tools leads to Resnick's second difference between learning in school and out of school: *pure mentation versus tool manipulation.* In schools, the greatest premium is often placed on pure thought activities. What can individuals do without the external support of books, notes, calculators and other instruments? Although these tools are used during learning activity, they are almost always absent during testing and evaluation. Thus, implicitly if not explicitly, schools value thought that could proceed independently without the aid of physical and cognitive tools. In contrast, most mental activities outside of school have an intimate connection to knowledge tools, and so the cognitive activity is shaped and dependant upon the kinds of tools that are available. For example, the shared knowledge of navigation has been shaped and impacted by the tools of navigation, progressing from an observation of the stars to magnetic compasses to gyrocompasses to global positioning and satellite technology.

A third difference between school and outside of school learning is *symbol manipulation in school versus contextualized reasoning outside school.* This point is most easily understood by considering math. In school, children learn math computational skills by manipulating a variety of symbols according to the rules of mathematics. The symbols and rules are often learned through rote repetition and with little connection to real-life applications. While math may be the most direct explanation of this point, the study of spelling, grammar, punctuation, historical dates and geographic locations can also make extensive use of symbols and rules, leading Resnick to state, "The process of schooling seems to encourage the idea that the 'game of school' is to learn symbolic rules of various kinds, that there is not supposed to be much continuity between what one knows outside school and what one learns in school" (p. 15).

Contrasted with this is how most people operate outside school. Actions are intimately connected with objects and events encountered in everyday life, and people often use these objects and events directly in their reasoning. For example, Scribner (1984) documented how dairy warehouse workers used their knowledge of case size and physical space to do essentially math without using formal math symbols. One worker, describing how he filled an order for a half a case (a full case is 16), is quoted as saying, "I visualize it. I knew the case I was looking at had ten out of it, and I only wanted eight, so I just added two to it...I don't never count when I'm making the order..." (p. 26). Lave et al. (1984) provided another example from their work with a weight watchers group, documenting how a person measured out three-fourths of two-thirds of a cup by measuring out a two-thirds cup portion of cottage cheese and laying it on the counter, rounding it off into a pancake like form, dividing it into four quarters, and then sweeping away one quarter. In both cases, people were engaging in mental activity that made sense in terms of their result in a specific context; their actions were grounded in the logic and tools of their immediate situation.

The fourth difference identified by Resnick between in-school and out-of-school learning was stated *as generalized learning versus situation-specific competencies outside.* This difference seems less notable than the three already discussed. The purpose of formal schooling to a great extent is knowledge that can be transferred or generalized to other situations. While the situated cognition/knowledge/learning perspective suggests that situated learning is in fact how people learn to function in everyday life, Resnick also points out that situation-specific learning by itself can also be limiting. For example, if no generalized learning occurs, an individual may not know how to adapt to a change in tasks, steps or tools. Thus, the perspective is that a combination of generalized learning and situated learning is better than either one alone.

Storytelling

Storytelling as a conceptual construct comes from the idea of narrative thinking as the fundamental way people make sense of everyday happenings in their life and in the lives of others (Bruner, 1990). Meaning making is at the heart of narrative thinking and storytelling, therefore it is grounded in cultural psychology and socioculturalism. Gabriel (2000) notes storytelling has always been an art of the ordinary folk, adding that: (a) it has been the antitheses of psychological science up until the mid-20th century; (b) it has gained acceptance in the late 20th century as a legitimate area of scientific inquiry into the human condition, mostly through the work of a cultural anthropologist; and (c) stories are part of a sense-making process that can be researched *in situ.* Bruner (1990) argues that the cognitive revolution of the 1950s started out as an effort to increase understanding of the role of the mind in the human sciences and that its aim "was to discover and to describe formally the meanings that human beings created out of their encounters with the world…" (p. 2). With the invention of the computer and conceptualization of the mind as an information processor, cognitive psychology veered sharply from its original aims (Bruner, 1990).

Two books, *Narrative Psychology: The Storied Nature of Human Conduct*, edited by T.R. Sarbin (1986) and *Acts of Meaning* by J. Bruner (1990) provide detailed explanations of research foundations of storytelling from the perspective of narrative psychology. H.P. Cole, in his 1997 article, *Stories to Live By*, provides a brief but cogent explanation of narrative thinking and storytelling. All three sources help explain why and how narrative thinking may provide a better frame of reference for understanding human behavior and ultimately influencing it. Several statements are presented below that represent major principles of narrative thinking and storytelling. These are taken or summarized from the three sources identified above. Their purpose is to provide readers with a quick but robust understanding of narrative thinking and storytelling as it relates to human thinking and behavior.

- There are two modes of cognitive thought: paradigmatic and narrative. Paradigmatic thinking is concerned with the construction of context-free, abstract, and formal concepts and principles. It is the domain of science, logic and mathematics. Narrative thinking is concerned with explanations of everyday life. It is the domain of ordinary people in everyday life and is also basic to scientific inquiry.
- Narrative thought is indigenous to human beings. The ability for meaning making exists in infants via their innate capacity to form mental representations of environmental phenomena. Babies and toddlers show evidence of an understanding of cause and effect relationships—and therefore meaning—well before they acquire formal language skill. Innate mental representations are triggered by the acts and expressions of others and by certain basic social contexts in which human beings interact.
- The acquisition of language occurs more readily through *use* rather than just exposure to it. By *using* language in the midst of *doing*, children learn not only what to say, but also how, where, to whom and under what circumstances. Therefore, language itself is *contextualized*. The situation is the same for adults learning a new language in a new country.
- Language and other knowledge tools are used to form and express individual knowledge, and to share knowledge with others. All knowledge is contextualized, or situated. It is the sharing (narrating, storytelling) of situated knowledge that creates common values and beliefs among like-minded groups of people. Values are inherent in commitment to "ways of life," and ways of life constitute a culture.
- It is in reflecting upon experiences or past actions of ourselves and of others that people construct stories. Thus storytelling is a type of causal thinking. Storytelling has structure. It involves actors and actions, a beginning, sequential sentences, and an ending. The functional purpose of its structure is to explain actions, events and happenings.
- Narrative thinking as a process is the translation of one's own and others' experiences into stories that integrate facts, perceptions, emotions, intentions, actions and consequences into coherent meaning.
- As children attempt to make meaning of their own and others' experiences, they readily construct and apply (imagine) missing elements (characters, actions, facts, reasons, etc.) so that a story makes sense to them. Older children and adults also supply missing elements in storytelling but rely more on their own personal situational experiences to fill in missing elements.
- All humans are subject to knowledge, reasoning, motivation, attention and memory limitations. For this reason, the details and "minutia" of situational experience are often not remembered accurately. The inaccurate recall of actual experience, combined with the human tendency to fill in missing story elements, is why narrative thinking (storytelling) is referred to as inherently subjective.
- Every important situational experience adds to a person's collection of situated cognitions. The subjective nature of narrative thinking, and the dynamic nature of situational experience and cognitions, combines to cause a person to constantly revaluate explanations of their own experiences. A result of this is that explanations of particular events or happenings are not consistent over time.

In summary, storytelling is a way of constructing meaning and understanding of one's own and others' experiences in narrative form. It is not the only way of organizing cognitive processes (e.g., perception, thought, action, memory), but it is a particularly relevant way in its natural domain of everyday life (Lincoln & Guba, 1986). The robustness and complexity of life events are better captured and communicated through narrative stories than through abstract, paradigmatic thinking (Cole, 1997).

Enculturalization

Enculturalization refers to the process by which a person adapts to a culture and assimilates its values. It is not mentioned specifically in Cole's (1997) conceptual model, but the notion of enculturalization bridges situated learning and cultural thinking and behavior. It is harmonious with Cole's model and a brief mention of it supports the idea of narrative thinking and storytelling as an important means of transmitting cultural values and beliefs.

In simple terms, enculturalization is what people do when learning to speak, read, write, or learn to become school children, office workers, professors and safety specialists. Brown et al. (1989) provides a good overview of enculturalization:

> From a very early age and throughout their lives, people, consciously or unconsciously, adopt the behavior and belief systems of new social groups.

> Given the chance to observe and practice *in situ* the behavior of members of a culture, people pick up relevant jargon, imitate behavior, and gradually start to act in accordance with its norms. These cultural practices are often recondite and extremely complex. Nonetheless, given the opportunity to observe and practice them, people adopt them with great success (p. 24).

Enculturalization takes place primarily through language as it provides a process for sorting out and communicating thoughts about the world. Language is the means by which stories are told and it is through stories that much enculturalization takes place. The study of how young children develop language skills—narrative thinking, stories—and learn to communicate provides important connecting links between later situated cognitions and narrative thinking. The intricacy of children's language development goes well beyond the scope of this review (see Berman and Slobin, 1994 for a comprehensive review of this topic). For our purposes, it is necessary only to understand that cultural values and beliefs, and the behaviors and explanations that follow, as expressed through stories, are passed from one generation to another. This enculturalization is a significant part of the factors that contribute to farming's hazardness and difficulty in reducing hazards and risk.

Empowerment Education

Empowerment theory comes from the work of Brazilian educator Paulo Freire (1921-1997) and his work in literacy with the poor of northeastern Brazil. Freire's philosophy of education first appeared in English in 1971 (Freire, 1971). Freire had two major premises that are relevant to narrative psychology. First, it is man's ontological vocation to act upon the world to transform it into ever new possibilities of fuller and richer life individually and collectively. Second, Freire viewed education as inherently nonneutral. As described by Shaull:

> There is no such thing as a neutral educational process. Education either functions as an instrument which is used to facilitate the integration of the younger generation into the logic of the present system and bring about conforming to it, *or* it becomes the "practice of freedom," the means by which men and women deal critically and creatively with reality and discover how to participate in the transformation of their world (in Freire, 1971, p. 15).

Freire proposed realistic problem solving and group dialogue as the effective way to educate and empower workers and individuals. Wallerstein and Bernstein (1988), using Freire's ideas, proposed a theory of "Empowerment Education" for health education. Wallerstein and Bernstein defined empowerment as, "a social action process that promotes participation of people, organizations, and communities in gaining control over their lives in their community and larger society" (p. 380). Wallerstein and Bernstein referenced several points from Freire in their article and these are summarized as:

- Empowerment...involves group efforts to identify their problems, to critically assess social and historical roots of problems, ...develop strategies to overcome obstacles in achieving goals.
- The goal of group dialogue is critical thinking by posing problems in such a way as to have participants uncover root causes of their place in society—the socioeconomic, political, cultural, and historical context of personal lives.
- Emphasis is on the collective knowledge that emerges from a group sharing experiences and understanding the social influences that affect individuals' lives.
- Emphasizes action and subsequent reflection as keys to the learning process.

Wallerstein & Weinger (1992) have applied empowerment education to worker safety and health. In this article, empowerment education is summed up as "an approach to learning that is participatory, is based on real-life experience, incorporates dialogue between and among educators and workers, critically analyzes the organizational and system-wide causes for problems, and has the goals of worker action and empowerment" (p. 620). Wallerstein and Weinger go on to compare these principles of empowerment with basic tenets from adult education theory. They note that:

- People retain information best when they are actively involved in problem-solving exercises and hands-on learning.... Education is less effective when people passively receive information, such as through a lecture or didactic slide presentation.
- Education is most effective when it includes the context of behaviors, including an analysis of obstacles to safe work practices. It is not effective if the focus becomes narrowed to teaching specific behaviors or competencies (p. 620).

Conceptually, empowerment education is easily understood and is supported by a variety of research from the social psychological tradition. Modern applications of *action research*, first proposed by Kurt Lewin (1890-1947) (Cook & Koss, 2001), and constructs such as *good self-esteem*, *positive self-concept*, *self-efficacy* and *locus of control* (Judge et al., 1998) relate with the central ideas of empowerment education. Wallerstein (1992) reviews a wide range of social and psychological studies from the perspective of empowerment theory and notes, "Through challenging social and psychological risk factors in a collective setting, people gain a belief they can control their worlds, a sense of their commonality, an ability to work together to acquire resources, and an actual transformation of socio-political conditions" (p. 202).

Luskin et al. (1992) refer to empowerment education as "learning-centered and participatory." They note that a major goal of this approach to worker safety and health was for "students to take with them the fundamental belief that they had the ultimate responsibility for their own safety and health" (p. 667). Weinger and Lyons (1992) used "participatory and action-oriented" educational methods in developing and implementing two farm worker pesticide safety programs in Nicaragua and in the United States and Puerto Rico. The authors stated, "The success of the popular education approach...demonstrates its adaptability to a variety of work settings and target groups" (p. 688). The principles of empowerment education also are well established in typical industrial safety and health training texts (e.g., Miller, 1998 and Hilyer et al., 2000).

Cole (1997) has suggested simulation exercises as an effective way to empower workers in safety and health education and provides a review of supportive research. In the same article, he has also explained how simulation exercises are compatible with narrative storytelling. Cole, along with his mine safety and health colleagues, has published a number of studies involving simulation exercises to improve miner safety and health performance (e.g., Cole et al., 1988; Cole, 1991; Cole et al., 1993). Cole and his colleagues in agricultural safety and health have applied simulation exercise principles to agricultural safety (Cole et al., 1997; Kidd et al., 1998). Their "Kayle's Difficult Decisions" (KDD) simulation exercise depicts the struggles of a typical farm family as they try to balance current work load and debt with possible expansion (Britt et al., 1999). Evaluation of the simulation exercise suggests that participants expect to make better production and safety decisions, better anticipate and plan work loads, and help make safety and injury prevention a higher priority on the farm.

Stories heard, stories lived and stories told (Howard, 1991) offer a methodology for capturing the intricate influences of all the factors that contribute to farming's hazardness (Table 4.1). Reality in agricultural safety and health may not reside in cognitive and behavioral theories and models that reduce human thinking and behavior into discrete and ordered bits to be coded and reassembled in various forms of values, beliefs, attitudes, motivations, goals, intentions, consequences and feedback loops. It may not be found in the analyses and interpretations of farm injury data. The reality may be in what individuals say their reasons are for the specific things they do in a specific situation, and in the issues that underlie the *farm safety—risk paradox*. This reality is more effectively captured orally and *in situ* in small group settings than by a typical printed questionnaire or telephone interview collected individually.

Chapter Summary

In Chapter 4, Table 4.1 identified several environmental, personal, work and social, economic and political factors that contribute to a culture of hazard and risk acceptance in agriculture. Farm operators, youth and families routinely express *concern* about hazards and risk when they talk about farm safety, but also they express *resignation* to accepting these hazards and risks as inherent to their way of life. It is suggested here that the key link between their concern and resignation, and farm safety and health education is *conversation.*

A point made in the previous chapter bears repeating: the ultimate goal of safety and health education is behavior change. Behavior is influenced by multiple factors that are biological, psychological, cultural and environmental (Wandersman, 2001). It is the intricacies of interconnected influence factors within each individual that directs a person's at-risk or safe behavior at any point in time. It may be that the closest safety and health professionals can hope to come to really understanding the intricacies that influence an individual's or a group's at-risk or safe behavior at a particular point in time is through the words and explanations of these individuals and groups relating their life activities, goals and plights. This is the essence of narrative psychology, storytelling and focus groups.

The truest test of a theoretical model's value in guiding safety and health education is its ability to generate structured interventions that can be experimentally evaluated. The newness of the Cole model means that its utility as a functional guide for educators is only beginning, but in combination with empowerment education (e.g., education that is participatory, is based on real-life experiences, involves problem solving), it seems to offer significant possibilities for farm safety and health education.

The final chapter of this review brings together thoughts and ideas for improving agricultural safety and health professional practices that do not easily fall within specific chapters.

7 Challenges, Opportunities, and Ideas for the Future

In the first chapter I mentioned that this review was written for agricultural injury prevention professionals, and that one advantage of a single-authored document over a consensus document was that it could dig deeper into sensitive issues and be more provocative and philosophical. At the same time, it is incumbent upon an author conducting a critical review of a field of practice to offer suggestions and recommendations for the future. That is the purpose of this chapter.

This review has been rather wide ranging, and is admittedly uneven in its treatment of some topics. For example, considerable space was devoted to the background and professional development of early agricultural safety professionals, the history of public health involvement in agricultural safety and health, and the BBS process. There has been little to no published literature in agriculture on these topics. On the other hand, some subjects, like behavioral and learning theory, are nearly bottomless wells in which one can spend months tracing the roots of a single philosophical viewpoint. Several times in the chapters addressing behavioral and learning theory, a statement was made that going further into a particular point was beyond the scope or purpose of this review. Nevertheless, an attempt was made to include enough insight into the topic to illustrate relevance to the broad issues of agricultural safety and health. References were provided if readers chose to explore the point or issue further. Finally, some more traditional topics in agricultural safety (e.g., safety engineering, human factors engineering [ergonomics], pesticide safety) were barely mentioned due to time limitations and a more complete existing literature on those topics.

Looking Toward the Future

All of the topics and subjects addressed in this review have a bearing on future directions and accomplishment for those working professionally for improved agricultural safety and health. While several problematic issues, questions and points of discord have been raised, there are few simple answers or solutions to be offered, and only a few have been suggested thus far (e.g., Table 4.2). The nature of the farming enterprise, the reluctance society has toward coerced interference with safety-related behavior on the farm, and the instability of resources and direction inherent with voluntary and public appropriation of funds, preclude quick fixes or easy resolutions. However, some opportunities abound.

The fact that these problems and issues present formidable challenges to agricultural safety and health professionals is nothing new: it has always been this way. As Chapman (2000) recently wrote, "Agricultural injury is not alone as an area where there are long dark tunnels with precious little light" (p. 176). Earlier in this review the point was made that agricultural safety and health has been mostly compassion driven, rather than theory or evaluation driven. It is suggested now that compassion must remain an equal partner with theory and evaluation, for compassion is our motivation to see the light at the end of the tunnel for moving out of darkness rather than the headlight of an oncoming train of endless injury.

Most injury prevention professionals readily understand that product, workplace and environmental modification are more effective and efficient strategies for a lasting reduction in work risk and injury than is a reliance on individual behavior. Yet what an individual does—his or her behavior—is often the determining factor in whether or not a *person* becomes injured at a specific point in time. This is why so much energy has been spent trying to understand, predict and change human behavior. With great difficulty, this was discussed in Chapters 5 and 6, and yet another new approach was identified and promoted. The opportunity before us is to take advantage of all the approaches of which we know, and to utilize them to more effectively interact with those who live and work in agriculture. To the extent that we can help agricultural workers understand the intricacies of their own safety-related decisions and behaviors, and the limitations inherent in being human, the more effective educational strategies can be used to their fullest possible extent.

There is light beyond the tunnel. The main professional organization of agricultural safety and health, NIFS, Inc., has become more diversified, primarily with the influx of professionals with public health-based backgrounds. This has helped to broaden the view of many with experience and professional training only in agriculture, while providing those with less experience and familiarity with agriculture with the opportunity to gain better insight. There has been and continues to be a considerable number of cooperative ventures between traditional agriculturalists and public health-based professionals. In fact, few if any major national initiatives or professional development efforts are undertaken today without significant involvement by both groups. Theory-driven interventions and intervention evaluation studies are beginning to find their way into literature. And new resources are available to more effectively guide occupational safety and health intervention evaluation during the next decade. There has never been more consistent, widespread and public support for the safety of youth living and working on farms. The National Initiative in Agricultural Safety and Health that began in 1990 has created an unprecedented level of research, program resources, and professional expertise that can be brought to the table for improving agricultural safety and health.

A new report, *Safe Work in the 21st Century: Education and Training Needs for the Next Decade's Occupational Safety and Health Personnel* has recently been released by the Institute of Medicine (2000), a division of the National Academy of Sciences. A committee of twelve experts, with extensive credentials and a variety of experiences, developed the report. Most interestingly, one committee member raised a point of contention about the committee makeup. His point was that the committee makeup did not sufficiently represent the views of all those who had a stake in the report's contents. This is similar to Field's (2000) point when discussing his experience during a visit to a farmer who had lost both arms to an unguarded PTO shaft. A question by the farmer about what he "really wanted" caused Field to reflect upon his own listening and seeing skills, and upon those who professionally practice in the field of agricultural safety and health. If you have not read this article, you should.

It is in the spirit of constructive reflection that this review has been conducted, and suggestions and recommendations are offered for consideration and debate by all of those interested in the professional practice of agricultural safety and health. The suggestions and recommendations listed below stem largely—but not exclusively—from the discussions in the latter part of Chapter 3 (The National Initiative in Agricultural Safety and Health) and Chapter 4, but do not necessarily connect one for one with the problems, challenges and issues identified in those or other chapters. Rather, the strategy taken with regard to suggestions and recommendations was to consider a possible systems approach for addressing the many interrelated problems, challenges and issues raised.

This approach was taken because the problems and issues identified throughout the chapters are so closely connected that what happens in one area of concern directly impacts one or more other areas of concern. For example, discussions and decisions regarding improving injury data quality (i.e., making it more consistent and useful) are closely connected to the issue of data for surveillance verses etiologic research purposes and the issue of distribution of funds by NIOSH to extramural verses intramural programs. This in turn significantly impacts how extramural dollars may be directed, which may then significantly impact the nature of injury prevention programs. For instance, taking advantage of methods of safety education that incorporate contemporary learning theory will require development of a number of new resource materials and training workshops for state and county farm safety leaders.

The result of this approach and process is a series of suggestions and recommendations that will hopefully contribute to a new national dialogue on improving agricultural safety and health. It will be in the process of a national dialogue that the more specific problems and concerns identified in this review will be addressed. Ideally, this new national dialogue would engage all stakeholders in significant ways and have an importance, visibility and impact that compares favorably to the 1991 Surgeon General's Conference on Agricultural Safety and Health. That Surgeon General's Conference proposed a national vision and agenda for production agriculture safety and health. It is time to revisit this national vision and agenda, particularly with the advantages of the experience and lessons learned during the past 10 years. At the same time, it seems appropriate to suggest that some party, organization or agency (or a combination) with some distance between current programs and initiatives should ideally be the one(s) to facilitate this process.

The suggestions and recommendations that follow are more general than detailed, a few are perhaps more idealistic than practical, but they are all in keeping with the idea of a national vision and agenda. The first suggestion is a call for a single national plan for production agriculture safety and health, with two ideas for how this might be approached. Suggestions and recommendations two through nine focus more directly on the problems, challenges and issues raised throughout this review, primarily those in the latter part of Chapter 3 and Chapter 4. Should the first suggestion be acted upon, it is likely that many of the concerns and issues represented in two through nine would be addressed. These could, however, be acted upon independently.

Suggestions and Recommendations

1. There should be a single national agenda for agricultural safety and health. National farm service and commodity group policy representatives (e.g., American Farm Bureau, National Pork Producers Council), U.S. legislators, federal agencies with occupational safety and health responsibility, and other stakeholder groups (cooperative extension, the equipment industry, state farm organizations, farm parents, teen workers, agricultural employers, migrant and seasonal farmworkers, etc.,) must be involved in this process from the beginning.
   - This national agenda could be developed around the concept of national strategic plans for surveillance, research, interventions, policy development, professional development, and information dissemination. The strategic plans should include short-term (3-5 year) and long-term (10-15 year) goals, objectives, and evaluation metrics.

- Only after a national agenda and strategic plans are developed and publicized, should a national effort be undertaken to identify and coordinate roles, priorities, funding, and implementation strategies for and by major stakeholder groups.

2. A consensus process is needed for guidelines on how dollars for the National Initiative should be divided between surveillance, research, intervention, professional development, and information dissemination efforts. Major questions behind this consensus process would be the most effective use of the National Initiative intramural and extramural funds, the Ag Centers concept, and support for cooperative extension and/or other rural community-based systems for helping to achieve Healthy People 2010 goals.
3. Alternative structures for the Ag Centers concept should be explored. For example, as opposed to the multi-function structure that characterizes present Ag Centers, Ag Centers structured by a target audience (e.g., family farm, migrant and seasonal workers, industrialized operations, etc.) or major program function (e.g., surveillance and data, engineering and ergonomics research, education/promotion, professional development, etc.) may be possible.
4. A consistent, inclusive and ongoing agricultural work injury surveillance program that provides reliable and valid estimates of injury rates by exposure hours at the national and state level is needed, along with timely dissemination of such data.
5. The agricultural safety and health profession should have available a more fully defined, developed, and supported set of competencies for its areas of practice. At a minimum this should include competencies in broad areas such as research, epidemiology, surveillance, education/promotion, and intervention evaluation, and be inclusive of factors such as those associated with the nature of farming and the *farm safety—risk paradox*. Avenues to achieve these prescribed competencies should include academic course work, experiential learning opportunities, continuing professional development, and professional certification (e.g., CSP, CIH, PE).
6. Irrespective of specific professional competencies, professional development for practitioners is needed around the issues of surveillance and research, injury prevention methods, intervention evaluation effectiveness methods, behavioral and learning theory and principles, developmentally appropriate tasks for children who work on farms, and ergonomic solutions to workplace hazards and environments.
7. A consensus document to establish a nomenclature is needed for agriculture injury data. At a minimum, this should include categories for age, source of injury, activity at time of injury, and for terms describing exposures, numerators, and denominators.
8. A major effort is needed to gain national grass-roots support from farm parents, farm organizations, and rural communities for minimal, national standards of acceptable conduct for deliberate exposure of children under the age of 16 to farm work and farm work environment hazards.
9. The production agriculture safety and health movement could benefit from a process and forum that would allow experienced injury prevention professionals an opportunity for in-depth exploration of the cultural, social and political barriers and impediments to improved agricultural safety and health.

In Chapter 1, the implication was made that a casual look at agricultural safety and health achievements, barriers, challenges and opportunities can obscure or hide underlying influences and factors that need examination in broad daylight if we are to significantly reduce agricultural work-related injury. It has been my intention to uncover and illuminate some of these influences and factors. Should any of the thoughts and ideas expressed anywhere in this review begin or contribute to conversation or action that helps our profession more effectively reduce unintentional injury in the farm work environment, this review will have served its purpose.

References

Adams M [1997]. Rhythm of the seasons: A journey beyond loss. Marshalltown, IA: Sta-Kris, Inc.

Agricultural Safety and Health Committee of the American Conference of Governmental Industrial Hygienists [1980]. Solomon's baby – the farmer… (editorial). Am Ind Hyg Assoc J, *41*: A-4.

Aherin RA [1987]. The prediction and analysis of safety behaviors among dairy farmers in central Wisconsin [dissertation]. Minneapolis, MN: University of Minnesota, Department of Agricultural Engineering.

Aherin RA, Murphy DJ, Westaby JD [1992]. Reducing farm injuries: issues and methods. St. Joseph, MI: American Society of Agricultural Engineers.

Ajzen I [1985]. From intentions to actions: a theory of planned behavior. In: J Kuhl and J Beckmann, eds. Action—control: from cognition to behavior. Heidelberg: Springer, pp.11-39.

Ajzen I [1991]. The theory of planned behavior. Org Behav Hum Dec Proc, *50*: 179-211.

Ajzen I, Fishbein M [1980]. Understanding attitudes and predicting social behavior. Englewood Cliff, NJ: Prentice-Hall.

Ajzen I, Madden TJ [1986]. Prediction of goal-directed behavior: attitudes, intentions, and perceived behavioral control. J Exp Soc Psychol, *22*: 453-74.

Allen F [1998]. Health psychology: theory and practice. St. Leonards, NSW, Australia: Allen & Unwin Pty Ltd.

Aronson E [1999]. The social animal. 8th Ed. New York, NY: Worth Publishers.

Axelrod SJ [1949]. Health problems in industrialized agriculture. Am J Public Health, *39*(9): 1172-5.

Bandura A [1977]. Toward a unifying theory of behavioral change. Psychol Rev, *84*: 191-215.

Bandura A [1986]. Social foundations of thought and action: a social cognitive theory. Englewood Cliffs, NJ: Prentice-Hall.

Baron S, Estill CF, Steege A, Lalich N, eds. [2001]. Simple solutions: ergonomics for farm workers. Washington, DC: US Department of Health and Human Services, Public Health Services, Centers for Disease Control and Prevention, National Institute for Occupational Safety and Health, DHHS (NIOSH) Publication No. 2001-111.

Barone DF, Maddox JE, Snyder CR [1997]. Social cognitive psychology: history and current domains. New York: Plenum Press.

Basch CE, Sliepcevich EM, Gold RS, Duncan DF, Kolbe LJ [1985]. Avoiding type III errors in health education program evaluations: a case study. Health Educ Quart, *12*(4): 315-31.

Baumeister RF, ed. [1999]. The *self* in social psychology. Philadelphia, PA: Psychology Press.

Baxter Magolda MB [1992]. Knowing and reasoning in students: gender-related patterns in students' intellectual development. San Francisco: Jossey-Bass.

Bennett C [1975]. Up the hierarchy. J Extension, *13*(2): 7-12.

Bennett C [1979]. Analyzing impacts of extension programs. ESC-57. Washington, DC: US Department of Agriculture, Science & Education Administration.

Bereiter C [1994]. Constructivism, socioculturalism, and popper's world 3. Educ Res, *23*(7): 21-3.

Bereiter C [1997]. Situated cognition and how to overcome it. In: Kirshner D, Whitson JA, eds. Situated cognition: social, semiotic, and psychological perspectives. Mahwah, NJ: Lawrence Erlbaum Associates, pp. 281-300.

Berman RA, Slobin DI [1994]. Relating events in narrative: a crosslinguistic developmental study. Hillsdale, NJ: Lawrence Erlbaum Associates.

Berry CM [1959]. Occupational health programs for agriculture and other nonplant workers. Prepared for 1959 National Health Forum, Chicago, IL.

Berry CM [1971]. Rural employment. Am J of Public Health, *61*(12): 2474-6.

Bickman L [1987]. The functions of program theory. In: Beckman L, ed. Using program theory in evaluation. San Francisco, CA: Jossey-Bass Publishers, pp. 5-18.

Blair E [1999]. Behavior-based safety: myths, magic & reality. Prof Saf, *44*(8): 25-9.

Boer H, Seydel ER [1996]. Protection motivation theory. In: Conner M, Norman P, eds. Predicting Health Behaviour. Bristol, PA: Open University Press, pp. 95-120.

Borders AC [1920]. Squamous-cell epithelioma of the lip. J Am Med Assoc, *74*(10): 656-64.

Borders AC [1921]. Squamous-cell epithelioma of the skin. Ann Surg, *72*(2): 141-60.

Boud D, Keogh R, Walker D [1996]. Promoting reflection in learning: a model. In: Edwards R, Hanson A, Raggatt P, eds. Boundaries of adult learning. New York: Routledge, pp. 32-56.

Bower GH, Morrow DG [1990]. Mental models in narrative comprehension. Science, *247*: 44-8.

Brazelton RW, Fischer D, Knutson GD [1976]. A farm accident study in California. Special Publication 3066. Division of Agricultural Sciences, University of California Cooperative Extension, Davis Campus, Davis, CA.

Britt M, Chrislip D, Bayer S, Cole H, Kidd P, Parshall M, Isaccs S, Struttman T, Colligan M, Scharf T [1999]. Farm work planning simulation in multi-media: a comparative evaluation. Am J Ind Med, *1*(Suppl): 113-5.

Brown JS, Collins A, Duguid P [1989]. Situated cognition and the culture of learning. Educ Res, *18*(1): 32-42.

Browning SR, Truszczynska H, Reed D, McKnight RH [1998]. Agricultural injuries among older Kentucky farmers: the farm family health and hazard surveillance study. Am J Ind Med, *33*: 341-53.

Bruner JS [1986]. Actual minds, possible worlds. Cambridge, MA: Harvard University Press.

Bruner J [1990]. Acts of meaning. Cambridge, MA: Harvard University Press.

Bureau of Agricultural Economics [1942]. The prevention of accidents on farms and in homes. Washington, DC: US Department of Agriculture.

Bureau of Agricultural Economics [1948]. Accidents to farm people and farm workers, January-April 1948. Washington, DC: US Department of Agriculture.

Burke J [1987]. Historical overview of the agricultural safety movement. Technical Paper No. 87-5512. St. Joseph, MI: American Society of Agricultural Engineers.

Burns R [1995]. The adult learner at work: a comprehensive guide to the context, psychology and methods of learning for the workplace. Sydney, AU: Business & Professional Publishing.

Burns S [1996]. Artistry in training: thinking differently about the way you help people to learn. Warriewood, NSW, AUS: Woodslane Press.

Butler J [1974]. NIOSH agricultural safety and health task report. Cincinnati, OH: Department of Health, Education, and Welfare, Public Health Service, Center For Disease Control, National Institute for Occupational Safety and Health.

Butterfield BD [1990]. Children at work: a five-part series on child labor in the United States. The Boston Globe, April 22-26.

Cahill J [1996]. Psychosocial aspects of interventions in occupational safety and health. Am J Ind Med, *29*: 308-13.

Calanddruccio RA, Powers JH [1949]. Farm accidents: A clinical and statistical study covering twenty years. Am J Surg, *November*: 652-60.

Candy PC [1991]. Self-direction for lifelong learning: a comprehensive guide to theory and practice. San Francisco: Jossey-Bass Publishers.

Carrabba JJ, Field WE, Tormoehlen RL, Talbert BA [2000]. Effectiveness of the Indiana 4-H tractor program at instilling safe tractor operating behaviors and attitudes in youth. J Agr Saf Health, *6*(3): 179-89.

Castillo DN, Adekoya N, Myers JR [1999]. Fatal work-related injuries in the agricultural production and service sectors among youth in the United States. J Agromed, *6*(3): 27-41.

Castillo DN, Landen DD, Layne LA [1994]. Occupational injury deaths of 16- and 17-year-olds in the United States. Am J Public Health, *84*(4): 646-9.

CDC [1987]. Traumatic occupational fatalities – United States, 1980-1984. *MMWR, 36*: 461-70.

CDC [1999]. Framework for program evaluation in public health. *MMWR, 41*(RR-11): 1-40.

CDC [2001]. Centers for agricultural disease, injury research, education and prevention. RFA-OH-01-004. [http: //grants.nih.gov/grants/guide/rfa-files/RFA-OH-01-004.html]. Date accessed: 3/28/01.

Chapman L [2000]. Intervene more often, evaluate more carefully, build on what works. J Agr Saf Health, *6*(3): 175-7.

Chapman LJ, Scharf T [1994]. Knowledge gaps about stress-related disorders among agricultural workers. In Proceedings of the Symposium on Efforts to Prevent Injury and Disease Among Agricultural Workers. Washington, DC: US Department of Health and Human Services, Public Health Service, Centers for Disease Control and Prevention, National Institute for Occupational Safety and Health, DHHS (NIOSH) Publication No. 94-119: 237-245.

Chen HT [1994]. Theory-driven evaluations: need, difficulties, and options. Eval Practice, *15*(1): 79-82.

Christoffel T, Gallagher SS [1999]. Injury prevention and public health: practical knowledge, skills and strategies. Gaithersburg, MD: Aspen publishers, Inc.

Cole HP [1991]. Simulation exercises as part-time training devices in miner training programs. Proceedings of the Training Resources Applied to Mining XVIII Conference, Pennsylvania State University, Park, PA, August 12-14.

Cole HP [1997]. Stories to live by: a narrative approach to health behavior research and injury prevention. In: Gochman DS, ed. Handbook of health behavior research IV. New York: Plenum Press, pp. 325-49.

Cole HP [2000]. Knowledge is not enough (editorial). J Agr Saf Health, *6*(4): 245-7.

Cole HP [2001]. Cognitive-behavioral approaches to farm community safety education: a conceptual analysis. Paper presented at the Using the Past and present to Map Future Actions Conference, Baltimore, MD, March 2-3.

Cole HP, Kidd PS, Isaacs SG, Parshall M, Scharf T [1997]. Difficult decisions: a simulation that illustrates cost effectiveness of farm safety behaviors. J Agromed, *4*(1/2): 117-24.

Cole HP, Lineberry GT, Wala AM, Haley JV, Berger PK, Wasielewski RD [1993]. Simulation exercises for training and educating miners and mining engineers. Mining Eng, *44*(11): 1397-401.

Cole HP, Wasielewski RD, Lineberry GT, Wala A, Mallett LG, Haley JV, Lacefield WE, Berger PK [1988]. Miner and trainer responses to simulated mine emergencies. In: Mine safety education and training seminar. Washington, DC: US Bureau of Mines, IC 9185, Library of Congress Cat. No. TN 295.U4, pp. 56-77.

Committee on Trauma Research, Commission on Life Sciences, National Research Council and Institute of Medicine [1985]. Injury in America: A Continuing Public Health Problem. Washington, DC: National Academy Press.

Conner M, Sparks P [1996]. The theory of planned behaviour and health behaviours. In: Conner M, Norman P, eds. Predicting Health Behaviour. Bristol, PA: Open University Press, pp. 121-62.

Conrad KJ, Miller TQ [1987]. Measuring and testing program philosophy. In: Bickman L, ed. Using program theory in evaluation. Number 33. San Francisco, CA: Jossey-Bass Publishers, pp. 19-42.

Conroy C, Sciortino [1997]. Describing patterns of occupational agricultural deaths: the effects of case definition. J Saf Res, *28*(4): 273-81.

Cook SL, Koss MP [2001]. Action research: informing interventions in male violence against women. In: Schneiderman N, Speers MA, Silva JM, Tomes H, Gentry JH, eds. Integrating behavioral and social sciences with public health. Washington, DC: American Psychological Association, pp. 115-39.

Cordes DH, Rea DF [1988]. Health hazards of farming. Am Family Practice, *38*: 233-244.

Cranton P [1998]. Transformative learning: individual growth and development through critical reflection. In: Scott SM, Spencer B, Thomas AM, eds. Learning for life: Canadian readings in adult education. Toronto, Ontario: Thompson Educational Publishing, pp. 188-99.

Crawford JM, Wilkins JR, Mitchell GL, Moeschberger ML, Bean TL, Jones LA [1998]. A cross-sectional case control study of work-related injuries among Ohio farmers. Am J Ind Med, *34*: 588-99.

Creevey K [1942]. Farm accidents. NY State J Med, *42*: 2016-20.

Cummings PH [1991]. Farm accidents and injuries among farm families and workers. Am Assoc Occup Health Nurses J, *39*(9): 409-15.

Dannenberg AL, Fowler CJ [1998]. Evaluation of interventions to prevent injuries: an overview. Injury Prev, *4*: 141-7.

Darragh AR, Stallones L, Sample PL, Sweitzer K [1998]. Perceptions of farm hazards and personal safety behavior among adolescent farmworkers. J Agr Saf Health *Special Issue, 1*: 159-69.

Davis B, Sumara D, Luce-Kapler R [2000]. Engaging minds: learning and teaching in a complex world. Mahwah, NJ: Lawrence Erlbaum Associates.

Davis LH [1971]. USDA's safety program. Farm Saf Rev, *29*(6): 5-6.

Dean DL [1994]. How to use focus groups. In: Wholley JS, Hatry HP, Newcomer KE, eds. Handbook of practical program evaluation. San Francisco, CA: Jossey-Bass Publishers, pp. 338-49.

DeBarr KA [1993]. Predicting adolescents' behavioral intention regarding safe farm tractor operation [Dissertation]. Carbondale, IL: Southern Illinois University, Department of Health Education.

DeHaven H [1942]. Mechanical analysis of survival in falls from heights of fifty to one hundred fifty feet. War Med, *2*: 586-96.

DeJoy DM [1996]. Theoretical models of health behavior and workplace self-protective behavior. J Saf Res, *27*(2): 61-72.

DePasquale JP, Geller ES [1999]. Critical success factors for behavior-based safety: a study of twenty industry-wide applications. J Saf Res, *30*(4): 237-49.

DeRoo LA, Rautiainen RH [2000]. A systematic review of farm safety interventions. Am J Prev Med, *18*(4S): 51-62.

Dewar DM [1996]. Farm health and safety issues: do men and women differ in their perceptions? Am Assoc Occup Health Nurses J, *44*(8): 391-401.

Dewey J [1924]. Democracy and education: an introduction to the philosophy of education. Reprinted from 1916. New York: The Macmillan Company.

Division of Medical Sciences, National Research Council [1966]. Accidental death and disability: the neglected disease of modern society. Washington, DC: National Academy of Sciences.

Dominice P [2000]. Learning from our lives: using educational biographies with adults. San Francisco: Jossey-Bass.

Donham KJ [1990]. Prologue: agricultural occupational and environmental health: policy strategies for the future. Am J Ind Med, *18*: 107-19.

Donham KJ, Gustafson KE [1982]. Human occupational hazards from swine confinement. Ann Am Conf Gov Ind Hyg, *2*: 137-4442.

Donham KJ, Mutel CF [1982]. Agricultural medicine: the missing component of the rural health movement. J of Family Med, *14*(3): 511-520.

Donham KJ, Osterburg D, Meyers M, Lehtola C [1997]. Tractor risk abatement and control: the policy conference; final report. The University of Iowa, Iowa City, IA. September 10-12.

Dosman JA, Cotton DJ, eds. [1980]. Occupational pulmonary disease: focus on grain dust and health. New York: Academic Press.

Dosman JA, Crockcroft DW, eds. [1989]. Principles of health and safety in agriculture. Boca Raton, FL: CRC Press.

Doss HJ, Pfister RG [1972]. Nature and extent of use of farm machinery in relation to frequency of accidents. Report No. 30. East Lansing, MI: Michigan State University, Center for Rural Manpower and Public Affairs.

Durso FT, ed. [1999]. Handbook of applied cognition. NY: John Wiley & Sons.

Eckenfelder DJ [1997]. It's the *culture*, stupid. Occup Haz, *59*: 41-4.

Elder JP, Apodaca JX, Parra-Medina D, Zuniga-Nuncio ML [1998]. Strategies for health education. In: Loue S, ed. Handbook of immigrant health. New York: Plenum Press, pp. 567-85.

Elkind PD [1993]. Correspondence between knowledge, attitudes and behavior in farm health and safety practices. J Saf Res, *24*: 171-9.

Elkind PD, Cody-Salter H [1994]. Farm stressors: the hazards of agrarian life. Ann Ag Environ Med, *1*: 23-7.

Erickson TA [1940]. Minnesota 4-H clubs. Talk given as part of a panel discussion on agricultural safety during the 29th National Safety Congress, Chicago, IL. October 10.

Estill CF, Tanaka S [1998]. Ergonomic considerations of manually harvesting Maine wild blueberries. J Agr Saf Health, *4*(1): 43-57.

Fetterman DM [1989]. Ethnography, step by step. Newbury Park, CA: SAGE Publications.

Field WE [2000]. ...Who have eyes but do not see...who have ears but do not hear (editorial). J Agr Saf Health, *6*(1): 11-2.

West Lafayette, IN: Purdue University Cooperative Extension Service

Finney JW, Moos RH [1989]. Theory and method in treatment evaluation. Eval Prog Planning, 12(4): 307-16.

Fishbein M, Ajzen I [1975]. Belief, attitude, intention and behavior: an introduction to theory and research. Reading MA: Addison-Wesley.

Fitch HG, Hermann J, Hopkins BL [1976]. Safe and unsafe behavior and its modification. J Occup Med, *18*: 618-22.

Freire P [1971]. Pedagogy of the oppressed. New York: Herder and Herder.

Fritsch CF [1976]. Occupational and nonoccupational fatalities on U.S. farms. Agricultural Economic Report No. 356, Economic Research Service, US Department of Agriculture, Washington, DC.

Fritsch CF, Zimmer JM [1980]. U.S. farm accidents, 1970-76. Agriculture Information Bulletin No. 434. Washington, DC: US Department of Agriculture; Economics, Statistics, and Cooperatives Service.

Gabriel Y [2000]. Storytelling in organizations: facts, fictions, and fantasies. New York: Oxford University Press.

Gaige M [1988]. The deadliest profession. Farm Journal, September, pp. 17-9.

Garnham A, Oakhill J [1994]. Thinking and reasoning. Cambridge, MA: Blackwell Publishers.

Gee JP [1997]. Thinking, learning, and reading: the situated sociocultural mind. In: Kirshner D, Whitson JA, eds. Situated cognition: social, semiotic, and psychological perspectives. Mahwah, NJ: Erlbaum Associates, Publishers, pp. 235-59.

Gelberg KH, Struttmann TW, London MA [1999]. A comparison of agricultural injuries between the young and elderly: New York and Kentucky. J Agr Saf Health, *5*(1): 73-81.

Geller ES [1996]. The psychology of safety: how to improve behaviors on the job. Radnor, PA: Chilton Book Company.

Gerberich SG, Gibson RW, French LR, Carr P, Renier CM, Gunderson PD, Martin F, True JA, Shutske J, Brademeyer K [1993]. The regional rural injury study-I (RRIS): a population-based effort. PB94134848. Springfield, VA: National Technical Information Service.

Gibson JJ [1961]. The contribution of experimental psychology to the formulation of the problem of Saf. In: Behavioral approaches to accident research. New York: Association for the Aid of Crippled Children, pp. 77-89.

Gilden RO [undated]. Extension service farm safety program. Committee report, 10 pp. Photocopy.

Gjerde C, Ferguson K, Mutel C, Donham K, Merchant J [1991]. Results of an educational intervention to improve the health knowledge, attitudes and self-reported behaviors of swine confinement workers. J Rural Health, *7*(3): 278-86.

Glasgow RE, Voit TM, Boyles SM [1999]. Evaluating the public health impact of health promotion interventions: the RE-AIM project. Am J Public Health, *89*(9): 1322-7.

Godfrey ES [1937]. Role of health department in the prevention of accidents. Am J Public Health, *27*: 152-5.

Goldenhar LM, LaMontagne AD, Katz T, Heaney C, Landsbergis P [2001]. The intervention research process in occupational safety and health: an overview from NORA intervention effectiveness research team. J. Occup Environ Medicine, *43*: 616-22.

Green KL [1999]. Farm health and safety: rural couples' beliefs and practices. J Agr Saf Health, *5*(1): 83-96.

Green LW, Kreuter MW [1991]. Health promotion planning: an educational and environmental approach. 2nd Ed. Mountain View, CA: Mayfield Publishing Co.

Green LW, Kreuter MW, Deeds SG, Partridge KB [1980]. Health education planning: a diagnostic approach. Palo Alto, CA: Mayfield Publishing Co.

Greeno JG [1993]. For research to reform education and cognitive science. In: Penner LA, Batsche GM, Knoff HM, Nelson DL, eds. The challenge in mathematics and science education: psychology's response. Washington, DC: American Psychological Association, pp. 153-92.

Greenwood MN [1983]. Guidelines for use of appropriated funds for state extension safety programs. Memo dated September 1983. Photocopy.

Grimaldi JV [1970]. The measurement of safety performance. J Saf Res, *2*: 137-59.

Guastello SJ [1993]. Do we really know how well our occupational accident programs work? Saf Science, *16*: 445-63.

Gunderson P, Donner R, Nashold L, Salkowicz [1993]. The epidemiology of suicide among farm residents or workers in five north-central states, 1980-1988. Am J Prev Med, *9* (3 Suppl.): 26-32.

Habermas J [1971]. Knowledge and human interest. Boston, MA: Beacon Press.

Haddon W Jr. [1963]. A note concerning accident theory and research with special reference to motor vehicle accidents. Ann NY Acad Sci, *107*: 636-46.

Haddon W Jr. [1970]. On the escape of tigers: an ecologic note. Technology Rev, *72*: 44-53.

Haddon W Jr. [1980]. Advances in the epidemiology of injuries as a basis for public health policy. Public Health Reports, 58: 1431-8.

Halpern DF [1996]. Thought and knowledge: an introduction to critical thinking. 3rd Ed. Mahwah, NJ: Lawrence Erlbaum Associates.
Halperin W, ed. [1992]. Public health surveillance. New York: Van Nostrand Reinhold, p xviii.
Hanford WD, Conrath D [1972]. Eight state survey: zeroing in on farm accidents. Farm Saf Rev, *30*(3): *Special Issue*.
Hanford WD, Burke JW, Fletcher WJ, Hoskin AF, Miller TA [1982]. 1982 farm accident survey report. Chicago, IL: National Safety Council.
Hanson A [1996]. The search for a separate theory of adult learning: does anyone really need andragogy? In: Edwards R, Hanson A, Raggatt P, eds. Boundaries of adult learning. New York: Routledge, pp. 99-108.
Hard DL, Myers JR, Snyder K, Casini V, Morton L, Cianfrocco R, Fields J [1999]. Young workers at risk when working in agricultural production. Am J Ind Med, (Suppl 1): 31-3.
Hawk C, Donham KJ, Gay J [1994]. Pediatric exposure to agricultural machinery: implications for primary prevention. J Agromed, *1*(1): 57-74.
Heinrich HW [1941]. Industrial accident prevention: a scientific approach. New York: McGraw-Hill.
Hessel ST, Strack J [1999]. National leadership in agricultural health: a history of the national farm medicine center. Marshfield, WI: National Farm Medicine Center.
Hettich P [1997]. Epistemological approaches to cognitive development in college students. In: Sutherland P, ed. Adult learning: a reader. Stirling, VA: Kogan Page Limited, pp. 46-57.
Heyer NJ, Franklin G, Rivara FP, Parker P, Haug JA [1992]. Occupational injuries among minors doing farm work in Washington state: 1986-89. Am J Public Health, *82*(4): 557-60.
Hickey P [1994]. Resistance to behaviorism. Behav Ther, *17*: 150-2.
Hidley JH [1998]. Critical success factors for behavior-based safety: avoiding common pitfalls & achieving real gains. Prof Saf, *43*(7): 30-4.
Hilyer BM, Veasey DA, Oldfield KW, McCormick LC [2000]. Effective safety and health training. Boca Ratan, FL: CRC Press.
Hineline PN [1994]. A self-interpretive behavior analysis. Am Psychol, *47*: 1274-1286.
Hodne CJ, Thu K, Donham KJ, Watson D, Roy N [1999]. Development of the farm safety and health beliefs scale. J Saf Health, *5*(4): 395-406.
Hoffman LE [1952]. Progress of agricultural colleges. Presentation at the 40th National Safety Congress, National Safety Council, Chicago, IL.
Hofmeister KM, Pfister RG [1968]. Michigan farm accident study: a report on accidents occurring to farm families and hired labor, 1967-68. Report No.14. East Lansing, MI: Michigan State University, Rural Manpower Center.
Hoskin AF, Miller TA, Hanford WD, Landes SR [1988]. Occupational injuries in agriculture: a 35-state summary. NIOSH Contract No. DSR-87-0942. Chicago, IL: National Safety Council.
Howard GS [1991]. Culture tales: a narrative approach to cross-cultural psychology and psychotherapy. Am Psychol, *46*: 187-97.
Hunt E [1999]. What is a theory of thought? In: Sternberg RJ, ed. The nature of cognition. Cambridge, MA: The MIT Press.
Hwang SA, Gomez MI, Stark AD, St. John TL, Pantea CI, Hallman EM, May JJ, Scofield SM [2000]. Safety awareness among New York farmers. Am J Ind Med, *38*: 71-81.
Ingersoll B [1989]. Perilous profession: farming is dangerous, but fatalistic farmers oppose safety laws. The Wall Street Journal, Midwest Edition, July 20, p. A1; p. A4.
Injury Epidemiology Division [1991]. Prevention of farm-related injuries in Oklahoma. Epidemiological Service. Oklahoma City, OK: Oklahoma State Department of Health.
Institute of Medicine [2000]. Safe work in the 21st century: education and training needs for the next decade's occupational safety and health personnel. Committee to Access Training Needs for Occupational Safety and Health Personnel in the United States, Board on Health Sciences Policy. Washington DC: National Academy Press.
Janicak CA [2000]. Occupational fatalities to workers age 65 and older involving tractors in the crops production agriculture industry. J Saf Res, *31*(3): 143-8.
Janz NK, Becker MH [1984]. The health belief model: a decade later. Health Educ Quart, 11: 1-47.
Jarvis P [1987a]. Meaningful and meaningless experience: towards an analysis of learning from life. Adult Educ Quart, *37*(3): 164-72.
Jarvis P [1987b]. Adult learning in the social context. London: Croom-Helm.
Jarvis P [1995]. Educating the adult educator in an information society: the role of the university. A paper presented at the International Conference on Educating the Educator: Role of the University, Canmore, Alberta.
Judge TA, Erez A, Bono JE [1998]. The power of being positive: the relation between positive self-concept and job performance. Hum Performance, *11*(2/3): 167-87.
Kagan J [1981]. The second year: the emergence of self-awareness. Cambridge, MA: Harvard University Press.
Keen J, Wallmer T [1989]. Danger on USA's farms: safety efforts take hold; risky business. USA Today, October 30, p. 1A; p. 8A.
Keifer MC [2000]. Effectiveness if interventions in reducing pesticide overexposure and poisonings. Am J Prev Med, *18*(4S): 80-9.
Kelly GA [1963]. The psychology of personal constructs. Vols. 1 & 2. New York: Norton.
Kelley WD, ed. [1982]. Agricultural respiratory hazards. Annals Am Conf Gov Ind Hygienists, Vol *2*, June.

Kelsey TK [1991]. Farm safety and federal responses to occupational health. J Rural Health, *7*(3): 287-92.

Kelsey TK [1994]. The agrarian myth and policy responses to farm safety. Am J Public Health, *84*(7): 1171-7.

Kennedy S [1995]. National occupational safety and health program in agriculture; Report of an external committee to review the extramural cooperative agreement programs. Submitted to Director, National Institute for Occupational Safety and Health, Centers for Disease Control and Prevention, Washington, DC.

Kidd PS, Isaacs SG, Cole HP, Parshall M, Scharf T, Struttmann T [1998]. An economic motivator for safe farming: changing perceptions through learning. J Agr Saf Health *Special Issue, 1*: 205-12.

Kidd P, Scharf T, Veazie M [1996]. Linking stress and injury in the farming environment: a secondary analysis of qualitative data. Health Educ Quart, *23*(2): 224-37.

Kimble GA [1996]. Psychology: a study of a science. Cambridge, MA: The MIT Press.

Kirby EL [1975]. Guidelines for use of FY 1975 funds for a state farm safety program. Washington, DC: US Department of Agriculture, Extension Service, dated January 22. Photocopy.

Kirshner D, Whitson JA, eds. [1997]. Situated cognition: social, semiotic, and psychological perspectives. Mahwah, NJ: Lawrence Erlbaum Associates, p. vii.

Knowles M [1970]. The modern practice of adult education: from pedagogy to andragogy. Chicago: Follett Publishing Company.

Knowles M [1973]. The adult learner: a neglected species. Houston, TX: Gulf Publishing Company.

Knowles M [1980]. The modern practice of adult education: from pedagogy to andragogy. Revised and updated. Chicago: Follett Publishing Company.

Knowles M [1984]. The adult learner: a neglected species. 3rd Ed. Houston, TX: Gulf Publishing Company.

Knudson T [1984]. A harvest of harm: the farm-health crisis. A series of articles, The Des Moines Register, September 16-23.

Kolb DA [1984]. Experiential learning: experience as the source of learning and development. Englewood Cliffs, NJ: Prentice-Hall.

Krause TR [1995]. Employee-driven systems for safe behavior: integrating behavioral and statistical methodologies. New York: Van Nostrand Reinhold.

Krause TR [1997]. The behavior-based process: managing involvement for an injury free culture. 2nd Ed. New York: Van Nostrand Reinhold.

Krause TR, Seymour KJ, Sloat KCM [1999]. Long-term evaluation of a behavior-based method for improving safety performance: a meta-analysis of 73 interrupted time-series replications. Saf Science, *32*: 1-18.

Krause TR, Hidley JH, Hodson SJ [1990]. The behavior-based safety process. New York: Van Nostrand Reinhold.

Landsittel DP, Murphy DJ, Kiernan NE, Hard DL, Kassab C [2001]. An evaluation of the effectiveness of educational interventions in the Pennsylvania central region farm safety pilot project. Am J Ind Med, *40*: 145-152.

Langley RL, McLymore RL, Meggs WJ, Roberson GT, eds. [1997]. Safety and health in agriculture, forestry, and fisheries. Rockville, MD: Government Institutes.

Last JM [1995]. A dictionary of epidemiology. 3rd Ed. New York: Oxford University Press.

Lave J, Murtaugh M, de la Roche O [1984]. The dialect of arithmetic in grocery shopping. In: Rogoff B, Lave J, eds. Everyday cognition: its development in social context. Cambridge, MA: Harvard University Press, pp. 67-94.

Lee BC [1995]. Factors influencing protection of children from agricultural hazards [Dissertation]. Milwaukee, WI: The University of Wisconsin-Milwaukee, School of Nursing.

Lee BC, Jenkins LS, Westaby JD [1997]. Factors influencing exposure of children to major hazards on family farms. J Rural Health, *13*(3): 206-15.

Lee BC [1998]. Alternatives to cognition. Mahwah, NJ: Lawrence Erlbaum Associates.

Lee TY, Gerberich SG, Gibson RW, Carr WP, Shutske J, CM Renier [1996]. A population-based study of tractor-related injuries: regional rural injury study-I (RRIS-I). J Occup Environ Health, *38*(8): 782-93.

Legault ML, Murphy DJ [2000]. Evaluation of the agricultural safety and health best management practices manual. J Agr Saf Health, *6*(2): 141-53.

Lewis MQ, Sprince NL, Burmeister LF, Whitten PS, Torner JC, Zwerling C [1998]. Work related injuries among Iowa farm operators: an analysis of the Iowa farm family health and hazard surveillance project. Am J Ind Med, *33*: 510-7.

Liller KD, Noland V, Lehtola CJ [2000]. An analysis of injury deaths on Florida farms for years 1989 through 1998. J Agr Saf Health, *6*(2): 131-40.

Lincoln YS, Guba EG [1986]. But is it rigorous? trustworthiness and authenticity in naturalistic evaluation. In: Williams DD, ed. Naturalistic evaluation: new directions for program evaluation. No. 30. San Francisco: Jossey-Bass, p. 73.

Lipsey MW [1996]. Key issues in intervention research: a program evaluation perspective. Am J Ind Med, *29*(4): 298-302.

Lipsey MW, Corday DS [2000]. Evaluation methods for social intervention. Annu Rev Psychol, 51: 345-75.

Lloyd EL [1933]. Agricultural accidents. Ind Med, *2*: 240-3.

Loafmann B [1997]. Feedback: the bridge from numbers to valuable information. Prof Saf, *42*(10): 38-9.

Locke EA [1991]. Goal theory vs. control theory: contrasting approaches to understanding work motivation. Motivation and Emotion, *15*(1): 9-28.

Locke EA, Latham GP [1990]. A theory of goal setting and task performance. Englewood Cliffs, NJ: Prentice-Hall.

Long HB [1998]. Understanding adult learners. In: Galbraith MW, ed. Adult learning methods: a guide for effective instruction. 2nd Ed. Malabar, FL: Krieger Publishing Company, pp. 21-35.

Lonner WJ, Adamopoulos J [1997]. Culture as antecedent to behavior. In: Berry JW, Poortinga YH, Pandey J, eds. Handbook of cross-cultural psychology: theory and method. 2nd Ed. Boston, MA: Allyn and Bacon, pp. 43-83.

Luskin J, Somers C, Woodring J, Levenstein C [1992]. Teaching health and safety: problems and possibilities for learner-centered training. Am J Ind Med, *22*: 665-76.

Madsen D, Klock LE, Wenzel FJ, Robbins JL, Schmidt CD [1976]. The prevalence of farmer's lung in an agricultural population. Am Rev Respir Dis, *113*: 171-4.

Mahoney MJ [1974]. Cognition and behavior modification. Cambridge, MA: Ballinger.

Mancuso JC [1986]. The acquisition and use of narrative grammar structure. In: Sarbin TR, ed. Narrative psychology: the storied nature of human conduct. New York: Praeger, pp. 91-110.

Mandel JH, Carr WP, Hillmer T, Leonard PR, Hallberg JU, Sanderson WT, Mandel JS [2000]. Safe handling of agriculture pesticides in Minnesota: results of a county-wide educational intervention. J Rural Health, *16*(2): 148-54.

Manstead ASR [1996]. Attitudes and behavior. In: Semin GR, Fiedler K, eds. Applied social psychology. Thousand Oaks, CA: SAGE Publications, pp. 3-29.

Marlenga B, Lee BC [1996]. Barriers to skin cancer screening follow-up in farmers. J Agromed, *3*(1): 5-16.

May S [1998]. Confronting issues affecting adult educators. In: Scott SM, Spencer B, Thomas AM, eds. Learning for life: Canadian readings in adult education. Toronto, Ontario: Thompson Educational Publishing, pp. 85-96.

Mayer RE [1996]. Learners as information processors: legacies and limitations of educational psychology's second metaphor. Educ Psychologist, *31*(3/4): 151-61.

McCurdy SA, Carroll DJ [2000]. Agricultural injury. Am J Ind Med, *38*: 463-80.

McDuffie HH, Dosman JA, Semchuk KM, Olenchock SA, Senthilselvan A, eds. [1995]. Agricultural health and safety: workplace, environment, sustainability. NY: Lewis Publishers.

McQueen DV [1996]. The search for theory in health behavior and health promotion. Health Promotion Quart, *11*(1): 27-32.

McSween TE [1995]. The values-based safety process: improving your safety culture with a behavioral approach. New York: Van Nostrand Reinhold.

Meier ST [1983]. Toward a theory of burnout. Hum Relations, *36*: 899-910.

Merchant JA, Kross B, Donham K, Pratt D, eds. [1989]. Agriculture at risk: a report to the nation. Iowa City, IA: The University of Iowa, Institute of Agricultural Medicine and Occupational Health, The National Coalition for Agricultural Safety and Health.

Merriam SB [1994]. Learning and life experience: the connection in adulthood. In: Sinnott JD, ed. Interdisciplinary handbook of adult learning. Westport, Ct: Greenwood Press, pp. 74-89.

Merriam SB, Caffarella RS [1999]. Learning in adulthood: a comprehensive guide. 2nd Ed. San Francisco: Jossey-Bass Publishers.

Merriam SB, Clark MC [1993]. Learning from life experience: what makes it significant? Int J Lifelong Educ, *12*(2): 129-38.

Mezirow J [1991]. Transformative dimensions in adult learning. San Francisco: Jossey-Bass Publishers.

Miller GA, Gildea PM [1987]. How children learn words. Scientific Am, *257*(3): 94-9.

Miller KL [1998]. Objective-based safety training. Boca Raton, FL: Lewis Publishers.

Mills EM, Shapiro M, Doss H [1989]. Farm injury in Michigan. Lansing, MI: Michigan Department of Public Health.

Mishler EG [1986]. The analysis of interview-narratives. In: Sarbin TR, ed. Narrative psychology: the storied nature of human conduct. New York: Praeger, pp. 233-55.

Mohler JC [1937]. What are outstanding causes of agricultural accidents? Proceedings of the 26th National Safety Congress, Chicago, IL. October 15.

Morse J, Field P [1995]. Qualitative research methods for health professionals. 2nd Ed. Thousand Oaks, CA: SAGE Publications.

Murphy DJ [1977]. 1976: A survey of farm accidents and illnesses in Pennsylvania. Extension Studies 70. University Park, PA: Penn State Cooperative Extension.

Murphy DJ [1992]. Safety and health for production agriculture. St. Joseph, MI: American Society of Agricultural Engineering.

Murphy DJ, Huizinga MA [1989]. A new approach to collecting farm accident data. J Saf Res, *20*: 21-9.

Murphy DJ, Senthilselvan A, Rennie D, Dosman J, Olenchock SA, eds. [2000]. J Agr Saf Health, *5*(1-4) (Compilation).

Murphy DJ, Yoder AM [1998]. Census of fatal occupational injury in the agricultural, forestry, and fishing industry. J Agr Saf Health S*pecial Issue, 1*: 55-66.

Murphy DJ, Kiernan NE, Chapman LJ [1996]. An occupational health and safety intervention research agenda for production agriculture: does safety education work? Am J Ind Med, *29*(4): 392-6.

Murphy DJ, Kiernan NE, Hard DL, Landsittel D [1998]. The Pennsylvania central region farm safety pilot project: part I—rationale and baseline results. J Agr Saf Health, *4*(1): 25-41.

Murphy DJ, Purschwitz M, Mahoney BS, Hoskin AF [1993]. A proposed classification code for farm and agricultural injuries. Am J Public Health, *83*(5): 736-8.

Murphy DJ, Selzer BA, Yesalis CE [1990]. Comparison of two methodologies to measure agricultural occupational fatalities. Am J Public Health, *80*(2): 198-200.

Myers JR [1990]. National surveillance of occupational fatalities in agriculture. Am J Ind Med, *18*(2): 163-18.

Myers JR [1998]. Injuries among farm workers in the United States, 1994. Washington, DC: US Department of Health and Human Services, Public Health Service, Centers for Disease Control and Prevention, National Institute for Occupational Safety and Health, DHHS (NIOSH) Publication No. 98-153.

Myers JR [2001]. Personal conversation on March 2 between JR Myers, Division of Safety Research, National Institute for Occupational Safety and Health, Centers for Disease Control and Prevention, Public Health Service, US Department of Health and Human Services, and DJ Murphy, Department of Agricultural and Biological Engineering, The Pennsylvania State University.

Myers JR, Hard DL [1995]. Work-related fatalities in the agricultural production and services sectors, 1980-89. Am J Ind Med, *27*: 51-63.

Myers JR, Hard DL, Snyder KA, Casini VJ, Cianfrocco R, Fields J, Morton L [1999]. Risks of fatal injuries to farm workers 55-years of age and older. Am J Ind Med *Supplement, 1*: 29-30.

Myers ML, Herrick RF, Olenchock SA, Myers JR, Parker JE, Hard DL, Wilson K, eds. [1992]. Papers and proceedings of the surgeon general's conference on agricultural safety and health. Cincinnati, OH: US Department of Heath and Human Services, Public Health Service, Centers for Disease Control, National Institute for Occupational Safety and Health, DHHS (NIOSH) Publication Number 92-105.

Napier TL, Goe WR, Pugh AR [1985]. Incidence and predictive factors associated with farm accidents in Ohio. Research Circular 287. Wooster, OH: Ohio State University, Ohio Agricultural Research and Development Center.

National Committee for Childhood Agricultural Injury Prevention [1996]. Children and agriculture: opportunities for safety and health. Marshfield, WI, Marshfield Clinic.

National Farm Medicine Center [1992]. National farm medicine center: past, present, future. Marshfield, WI: Author.

National Institute for Farm Safety, Inc. [1962]. Bylaws of the National Institute of Farm Saf, Inc.: Middleton Township, Dade County, WI. Photocopy.

National Institute for Farm Safety, Inc. [1973]. National Institute for Farm Safety, Inc.: Its People, Purpose, Progress. Madison, WI: Brochure published courtesy of Randall C Swanson and Edward S Adams.

National Institute for Farm Safety, Inc. [1982]. National Institute for Farm Safety, Inc. Madison, WI: Brochure published courtesy of Arthur F Stapel and Randall C Swanson.

National Safety Council [1937-38, 1940-43]. Proceedings of annual safety congresses: Agricultural safety sessions, 1937, 1938 and 1940; Farm safety sessions, 1941 and 1943; Home and farm safety session, 1942. Chicago, IL: Author.

National Safety Council [1943]. Sparkplugging farm safety. Farm Saf Rev, *1*(5): 2-3.

National Safety Council [1945]. Safe training in New York. Farm Saf Rev, *3*(2): 4-5, 15.

National Safety Council [1946]. Farm Saf Rev, *4*(2).

National Safety Council [1955]. Farm bureau directors met in St. Louis. Farm Saf Rev, 11: 15.
Photocopy.

National Safety Council [1971a]. Safety delegates meet with secretary Hardin. Farm Saf Rev, *29*(2): 7.

National Safety Council [1971b]. Accident facts: 1971 Ed. Chicago: Author.

National Safety Council [1981-2000]. Annual editions of injury facts (formerly accident facts). Itasca, IL: Author.

National Safety Council [2000a]. Injury Facts. Itasca, IL: Author.

National Safety Council [2000b]. Accident prevention manual for industrial operations. 12th Ed. Chicago: Author.

Neisser U [1967]. Cognitive psychology. New York: Appleton-Century-Crofts.

NIOSH [1996]. National occupational research agenda. Washington, DC: US Department of Health and Human Services, Public Health Service, Centers for Disease Control and Prevention, National Institute for Occupational Safety and Health, DHHS (NIOSH) Publication No. 96-115.

NIOSH [1997]. Agriculture: NIOSH research projects. Washington, DC: US Department of Health and Human Services, Centers for Disease Control and Prevention, National Institute for Occupational Safety and Health, DHHS (NIOSH) Publication No. 97-124.

NIOSH [1999a]. A model for research on training effectiveness. Washington, DC: US Department of Health and Human Services, Public Health Services, Centers for Disease Control and Prevention, National Institute for Occupational Safety and Health. DHHS (NIOSH) Publication No. 99-142.

NIOSH [1999b]. NIOSH childhood agricultural injury prevention initiative. [http: //www.cdc.gov/niosh/childagz .html]. Date accessed: November 2000.

NIOSH [2000]. Worker health chart book, 2000. Washington, DC: US Department of Health and Human Services, Public Health Service, Centers for Disease Control and Prevention, National Institute for Occupational Safety and Health, DHHS (NIOSH) Publication No. 2000-127.

NIOSH [2001a]. National traumatic occupational fatality (NTOF) database, 1980-1997. Washington, DC: US Department of Health and Human Services, Public Health Service, Centers for Disease Control and prevention, National Institute for Occupational Safety and Health. Unpublished database; accession date, May 9.

NIOSH. [2001b]. Tracking occupational injuries, illnesses, and hazards: the NIOSH surveillance strategic plan. Washington, DC: US Department of Health and Human Services, Public Health Service, Centers for Disease Control and Prevention, National Institute for Occupational Safety and Health, DHHS (NIOSH) Publication No. 2001-118.

NIOSH. [forthcoming]. Centers for agricultural disease and injury research, education and prevention: annual report summary fiscal year 2000. Washington, DC: US Department of Health and Human Services, Public Health Service, Centers for Disease Control and Prevention, National Institute for Occupational Safety and Health.

Nisbitt R, Ross L [1980]. Human inference: strategies and shortcomings of social judgment. Englewood Cliffs, NJ: Prentice-Hall.

Nisbitt RE, Wilson TD [1977]. Telling more than we can know: verbal reports on mental processes. Psychol Rev, 96: 341-57.

Norman GR [1999]. The adult learner: a mythical species. Academic Med, *74*(8): 886-9.

Norman P, Conner M [1996]. The role of social cognition in health behaviour. In: Conner M, Norman P, eds. Predicting Health Behaviour. Bristol, PA: Open University Press, pp. 1-22.

Norman P, Bennett P [1996]. Health locus of control. In: Conner M, Norman P, eds. Predicting health behaviour. Bristol, PA: Open University Press, pp. 62-94.

O'Donohue W, Kitchener R [1999]. Introduction: the behaviorisms. In O'Donohue W, Kitchener R, eds. Handbook of behaviorism. New York: Academic Press, pp. 1-13.

Office of Management and Budget [2000]. Circular A-21 (Revised), Subject Cost principles for educational institutions. http://www.whitehouse.gov/omb/circulars/a021/a021.html. Date accessed: September 2001.

Offutt S [2000]. Can the farm problem be resolved? M.E. John Lecture, Department of Agricultural Economics and Rural Sociology, The Pennsylvania State University, University Park, PA. October 18.

Oppegaard MO [1934]. Farm accidents. Minnesota Med, 17: 127-30.

Osorio AM, Geiser CR, Husting EL, Inai A, Summerill KF [1998]. Farm injury surveillance in two California counties—general findings. J Agr Saf Health Special Issue, 1: 99-108.

Parrott R, Lewis D, Jones K, Steiner C, Goldenhar L [1998]. Identifying feed and seed stores as a site to promote skin cancer control: a social marketing approach to agricultural health communication. J Agr Saf Health *Special Issue, 1*: 149-58.

Parrott R, Steiner C, Goldenhar L [1996]. Georgia's harvesting healthy habits: a formative evaluation. J Rural Health, *12*(4): 291-300.

Patterson KW, Novack JA, Betrand AL [1972]. The distribution and characteristics of farm accidents in Louisiana. Bulletin No. 665. Baton Rouge, LA: Louisiana State University, Agricultural and Mechanical College, Department of Rural Sociology.

Pederson DH, Wilkins JR III, Bean TL, Mitchell GL, Crawford JM, Jones LA [1999]. Agricultural hazard data from a population-based survey of cash grain farms: Ohio observations. Appl Occup Environ Hyg, *14*: 299-305.

Pennsylvania Department of Public Instruction, Commonwealth of Pennsylvania [1957]. A study of accidents to Pennsylvania farm people. Bulletin 398. Harrisburg, PA: Author.

Peterson D [1975]. Safety management: a human approach. Fairview, NJ: Aloray Inc.

Peterson D [1997]. Accountability, culture & behavior. Prof Saf, *42*(10): 45.

Peterson D [1998]. Safety management: a human approach. 2nd Ed. Fairview, NJ: Aloray Inc.

Petrea RE [1996]. Applying the theory of planned behavior to respiratory protection utilization and behaviors of east-central Illinois pork producers [Thesis]. Champaign, IL: University of Illinois at Urbana-Champaign, College of Education.

Pfister RG, Hofmeister KM [1969]. A study of accidents occurring to farm people in Michigan. J Saf Res, *1*(1): 28-41.

Phillips GH, Stuckey WE, Pugh AR [1975]. Farm family accidents: 1967 and 1972. J Saf Res, *7*(2): 85-9.

Powers JH [1939]. The hazards of farming. J Am Med Assoc, *113*: 1375-8.

Pratt DS [1990]. The Bassett farm safety and health project. Am J Ind Med, 18(4): 391-3.

Pratt DH, Marvel LH, Darrow D, Stallones L, May JJ, Jenkins P [1992]. The dangers of dairy farming: the injury experience of 600 workers followed for two years. Am J Ind Med, *21*: 637-50.

Prochaska JO, DiClemente CC [1983]. Stages and processes of self-change of smoking: toward an integrative model of change. J Consult Clin Psychol, *51*(3): 390-5.

Prochaska JO, DiClemente CC, Norcross JC [1992]. In search of how people change: applications to addictive behaviors. Am Psychologist, *47*(9): 1102-14.

Prochaska JO, Velicer WF, Rossi JS, Goldstein MG, Marcus BH, Rakowski W, Fiore C, Harlow LL, Redding CA, Rosenbloom D, Rossi SR [1994]. Stages of change and decisional balance for 12 problem behaviors. Health Psychol, *13*(1): 39-46.

Proctor RW, Weeks DJ [1990]. The goal of B.F. Skinner and behavioral analysis. NY: Springer-Verlog.

Purschwitz MA [1992]. Farm and agricultural injury statistics. In: Murphy DJ, Safety and health in production agriculture. St. Joseph, MI: American Society of Agricultural Engineers, p. 43.

Purschwitz MA [1997]. Epidemiology of agricultural injuries and illnesses. In: Langley RL, McLymore RL, Meggs WJ, Roberson GT, eds. Safety and health in agriculture, forestry, and fisheries. Rockville, MD: Government Institutes, p. 215.

Ragan PT [1997]. Safety's silver bullet? Prof Saf, *42*: 25-9.

Reber RA, Wallin JA [1983]. Validation of a behavioral measure of occupational safety. J Organizational Mgmt, *5*(2): 69-77.

Reber RA, Wallin JA, Chhokar JS [1984]. Reducing industrial accidents: a behavioral experience. Ind Relations, *23*(1): 119-25.

Rein BK [1998a]. Recommendations for extension farm safety funding. Memo dated January 28. Washington, DC: United States Department of Agriculture, Cooperative State Research, Education, and Extension Service. Photocopy.

Rein BK [1998b]. Guidelines for Extension Farm Safety Performance Plans. Memo dated March 16. Washington, DC: United States Department of Agriculture, Cooperative State Research, Education, and Extension Service. Photocopy.

Resnick LB [1987]. Learning in school and out. Educ Res, *16*(9): 13-20.

Rhodes DA, Hupcey JE [2000]. The perception of farm safety and prevention issues among the old order Amish in Lancaster County, Pennsylvania. J Agr Saf Health, *6*(3): 203-13.

Rivara FP [1997]. Fatal and nonfatal farm injuries to children and adolescents in the United States, 1990-1993. Injury Prev, *3*: 190-4.

Robertson LS [1983]. Injuries: causes, control strategies, and public health policy. Lexington, MA: Lexington Books, D.C. Heath & Company.

Robertson LS [1998]. Injury epidemiology. 2nd Edition. New York: Oxford University Press.

Robson LS, Shannon HS, Goldenhar LM, Hale AR [2001]. Guide to evaluating the effectiveness of strategies for preventing work injuries: how to show whether a safety intervention really works. Washington, DC: US Department of Health and Human Services, Public Health Service, Centers for Disease Control and prevention, National Institute for Occupational Safety and Health, DHHS (NIOSH) Publication No. 2001-119.

Rockwell K, Bennett C [2001]. Targeting outcomes of programs (TOP). [http: //deal.unl.edu/TOP/synopsis .html]. Date accessed: January 25.

Rosenblatt PC, Lasley P [1991]. Perspective on farm accident statistics. J Rural Health, *7*(1): 51-61.

Rogers RW [1975]. A protection motivation theory of fear appeals and attitude change. J Psychol, *91*: 93-114.

Rogers RW [1983]. Cognitive and physiological processes in fear appeals and attitude change: a revised theory of protection motivation. In: Cacioppo JT, Petty RE, eds. Social psychophysiology: a source book. New York: Guilford Press, pp. 153-76.

Rosenstock IM [1966]. Why people use health services. The Milbank Memorial Fund Quart, *44*(3): 94-124.

Rosenstock IM [1974]. Historical origins of the health belief model. Health Educ Monographs, *2*(4): 329-35.

Rossi PH, Freeman HE, Lipsey MW [1999]. Evaluation: a systematic approach. 6th Ed. Thousand Oaks, CA: SAGE Publications.

Rossiter M [1999]. A narrative approach to development: implications for adult education. Adult Educ Quart, *50*(1): 56-71.

Roth I [1990]. Challenging habits of expectation. In: Mezirow and Associates, eds. Fostering critical reflection in adulthood. San Francisco: Jossey-Bass Publishers.

Rotter JB [1954]. Social learning and clinical psychology. New York: Prentice-Hall.

Rotter JB [1966]. Generalized expectancies for internal and external control of reinforcement. Psychol Monographs: General and Applied, *80* (whole no. 609): 1-28.

Runyan JL [1998]. Injuries and fatalities on U.S. farms. Agricultural Information Bulletin Number 739. Washington, DC: US Department of Agriculture, Economic Research Service.

Runyan JL [2000]. Profile of hired farm workers, 1998 annual averages. Agricultural Economic Report Number 790. Washington, DC: US Department of Agriculture, Economic Research Service.

Ruser JW [1998]. Denominator choice in the calculation of workplace fatality rates. Am J Ind Med, *33*: 151-6.

Rush JD [1957]. Fatal farm accidents in the U.S. 1949-1953. ARS 43-50. Washington, DC: US Department of Agriculture, Agricultural Research Service.

Saari J, Nasanen M [1989]. The effect of positive feedback on industrial housekeeping and accidents: a long-term study at a shipyard. Int J Ind Ergo, *4*: 210-1.

Sarbin TR [1986]. The narrative as a root metaphor for psychology. In: TR Sarbin, ed. Narrative psychology; the storied nature of human conduct. New York: Praeger Scientific, pp. 3-21.

Savage EP [1974]. The Colorado view. Conference on Agricultural Safety and Health: Proceedings of Symposium. Society for Occupational and Environmental Health, Iowa City, IA. September 4-5, pp. 15-19.

Scharf T, Kidd P, Cole H, Bean T, Donham K, Baker D [1998]. Intervention tools for farmers—safe and productive work practices in a safer work environment. J Agr Saf Health *Special Issue, 1*: 193-203.

Scheirer MA [1994]. Designing and using process evaluation. In: Wholley JS, Hatry HP, Newcomer KE, eds. Handbook of practical program evaluation. San Francisco: Jossey-Bass Publishers, pp. 40-68.

Schenker MB [1996]. Preventive medicine and health promotion are overdue in the agricultural workplace. J Pub Health Policy, *17*(3): 275-304.

Schober MF [1999]. Making sense of questions: an interactional approach. In: Sirken MG, Hermann DJ, Schechter S, Schwarz N, Tanaur JM, Tourangeau R, eds. Cognition and survey research. New York: John Wiley & Sons, pp. 77-93.

Schulte PA, Goldenhar LM, Connally LB [1996]. Intervention research: science, skills and strategies. Am J Ind Med, *29*(4): 285-8.

Schwab CV, Ralston AR, Miller LJ, Hanna HM [1995]. Comparisons between perceptions of farm hazards and injury records in Iowa. J Agr Saf Health, *1*(4): 241-8.

Schwarz N [1999]. Cognitive research into survey measurement: its influence on survey methodology and cognitive theory. In: Sirken MG, Hermann DJ, Schechter S, Schwarz N, Tanaur JM, Tourangeau R, eds. Cognition and survey research. New York: John Wiley & Sons, pp. 65-75.

Schwarzer R, Fuchs R [1996]. Self-efficacy and health behaviours. In: Conner M, Norman P, eds. Predicting Health Behaviour. Bristol, PA: Open University Press, pp. 163-96.

Scott SM [1998]. Philosophies in action. In: Scott SM, Spencer B, Thomas AM, eds. Learning for life: Canadian readings in adult education. Toronto, Ontario: Thompson Educational Publishing, pp. 99-106.

Scribner S [1984]. Studying working intelligence. In: Rogoff B, Lave J, eds. Everyday cognition: its development in social context. Cambridge, MA: Harvard University Press, pp. 9-40.

Seagraves CM [1937]. A program for farm safety. Proceedings of the 26th National Safety Congress, Chicago, IL. October 15.

Sedikides C [1999]. Assessment, enhancement, and verification determinants of the self-evaluation process. In: Baumeister RF, ed. The *self* in social psychology. Philadelphia, PA: Psychology Press, pp. 402-25.

Seevers B, Graham D, Gamon J, Conklin N [1997]. Education through cooperative extension. Albany: NY: Delmar Publishers.

Semin GR, Fiedler K [1996]. Applied social psychology. Thousand Oaks, CA: SAGE Publications.

Shannon HS, Robson LS, Guastello SJ [1999]. Methodological criteria for evaluating occupational Saf intervention research. Saf Science, *31*: 161-79.

Sheeran P, Abraham C [1996]. The health belief model. In: Conner M, Norman P, eds. Predicting health behaviour. Bristol, PA: Open University Press, pp. 23-61.

Shum MS, Rips LJ [1999]. The respondent's confession: autobiographical memory in the context of surveys. In: Sirken MG, Hermann DJ, Schechter S, Schwarz N, Tanaur JM, Tourangeau R, eds. Cognition and survey research. New York: John Wiley & Sons, pp. 95-109.

Silverman SL, Casazza ME [2000]. Learning and development: making connections to enhance teaching. San Francisco: Jossey-Bass Publishers.

Sinnott JD, ed. [1994]. Interdisciplinary handbook of adult lifespan learning. Westport, CT: Greenwood Press.

Sirken MG, Schechter S [1999]. Interdisciplinary survey methods research. In: Sirken MG, Hermann DJ, Schechter S, Schwarz N, Tanaur JM, Tourangeau R, eds. Cognition and survey research. New York: John Wiley & Sons, pp. 1-10.

Skinner BF [1974]. About behaviorism. NY: Alfred A. Knopf.

Smilkstein R [1990-1991]. A natural teaching method based on learning theory. Gamut. Seattle Community College District, Seattle, WA, pp. 12-15, 36.

Smith PB, Bond MH [1999]. Social psychology across cultures. 2nd Ed. Boston, MA: Allyn and Bacon.

Smith MJ, Anger WK, Uslun SS [1978]. Behavior modification applied to occupational safety. J Saf Res, *10*: 87-8.

Smith TA [1999]. What's wrong with behavior-based safety? Prof Saf, *44*: 37-40.

Society for Occupational and Environmental Health [1974]. Conference on agricultural safety and health: proceedings of symposium. New York: SOEH.

Spigner-Littles D, Anderson CE [1999]. Constructivism: a paradigm for older learners. Educ Gerontology, *25*: 203-9.

Starnes BA [1999]. The foxfire approach to teaching and learning: John Dewey, experiential learning, and the core practices. ERIC Digest, EDO-RC-98-6. Charleston, WV: Appalachia Educational Laboratory.

Steers RM, Porter LW [1991]. Motivation and work behavior. 5th Ed. New York: McGraw-Hill, Inc.

Sternberg RJ [1999]. A dialectical basis for understanding the study of cognition. In: Sternberg RJ, ed. The nature of cognition. Cambridge, MA: The MIT Press, pp. 51-78.

Sternberg R, Wagner R, Williams W, Horvath J [1995]. Testing common sense. Am Psychologist, *50*(11): 912-27.

Stewart LE [1977]. Accidents on Maryland farms 1974. Bulletin 255. College Park, MD: University of Maryland Cooperative Extension Service.

Strobe W, de Wit J [1996]. Health-impairing behaviors. In: Semin GR, Fiedler K, eds. Applied social psychology. Thousand Oaks, CA: SAGE Publications, pp. 113-44.

Stuckey WE, Pugh AR [1973]. Accidental injuries to Ohio farm people, 1957-1972. Columbus, OH: Ohio State University Cooperative Extension Service.

Stueland D, Layde P, Lee BC [1991]. Agricultural injuries in children in central Wisconsin. J Trauma, *31*(1): 1503-9.

Stueland, D, Lee BC and P Layde [1991]. Surveillance of agricultural injuries in central Wisconsin: Epidemiologic characteristics. J Rural Health, *7*(1): 63-71.

Stueland D, Zock T, Stams Jr. P, Krieg G, Boulet W [1990]. The spectrum of emergency care of agricultural trauma in central Wisconsin. Am J Emerg Med, *8*(6): 528-30.

Stump RL [1979]. West Virginia farm accident survey. Morgantown WV: West Virginia University Cooperative Extension Service.

Sulzer-Azaroff B [1982]. Behavioral approaches to occupational safety and health. In: Frederiksen LE, ed. Handbook of organizational behavior management. New York: John Wiley & Sons, pp. 505-38.

Sulzer-Azaroff B, Fellner D [1984]. J Organizational Behav Mgmt, *6*(2): 53-65.

Sulzer-Azaroff B, Austin J [2000]. Does BBS work? behavior-based safety & injury reduction: a survey of the evidence. Prof Saf, *45*(7): 19-24.

Sulzer-Azaroff B, Lischeid WE [1999]. Assessing the quality of behavioral safety initiatives. Prof Saf, *44*(4): 31-6.

Sutherland P, ed. [1997]. Adult learning: a reader. Stirling, VA: Kogan Page Limited.

Sutton S [1997]. Transtheoritical model of behavior change. In Baum A, Newman S, Weinman J, West R, McManus C, eds. Cambridge Handbook of Psychology, Health and Medicine. NY: Cambridge University Press, pp. 180-83.

Taylor RB [1976]. Cultural ways: a concise edition of introduction to cultural anthropology. 2nd Ed. Boston, MA: Allyn & Bacon.

Teutsch SM [1992]. A framework for assessing the effectiveness of disease and injury prevention. *MMWR*, *41*(RR-3): 1-12.

Tevis C, Finck C [1989]. We kill too many farm kids. Successful Farming: Special Report, Mid-February, pp. 18A-18P.

Thacker SB. [2000]. Historical development. In: Teutsch SM, Churchill RE, eds. Principles and practice of public health surveillance. 2nd Ed. New York: Oxford University Press, p. 1.

Thu KM, Pies B, Roy N, Von Essen S, Donham K [1998]. A qualitative assessment of farmer responses to the certified safe farm concept in Iowa and Nebraska. J Agr Saf Health, *4*(3): 161-71.

Thu K, Donham KJ, Yoder D, Olgilve L [1990]. The farm family perception of occupational health: a multistate survey of knowledge, attitudes, behaviors and ideas. Am J Ind Med, *18*(4): 427-31.

Tight M [1996]. Key concepts in adult education and training. New York: Routledge.

Timmreck TC [1994]. An introduction to epidemiology. Boston, MA: Jones and Bartlett.

Tripp SD [1996]. Theories, traditions and situated learning. In: McLellan H, ed. Situated learning perspectives. Englewood Cliffs, NJ: Educational Technologies Publications, pp. 155-66.

Tofany VL [1978]. Appropriations concerning federal support for extension farm safety leaders. Memo dated February 21. Chicago, IL: National Safety Council.

Top FH [1956]. Preventive medicine in agriculture. J Am Med Assoc, *161*(14): 1357-60.

Torff B, Sternberg RJ [1998]. Changing mind, changing world: practical intelligence and tacit knowledge in adult learning. In: Smith MC, Pourchot T, eds. Adult learning and development: perspectives from educational psychology. Mahwah, NJ: Lawrence Erlbaum Associates, pp.109-25.

Tourangeau R [1999]. Context effects on answers to attitude questions. In: Sirken MG, Hermann DJ, Schechter S, Schwarz N, Tanaur JM, Tourangeau R, eds. Cognition and survey research. New York: John Wiley & Sons, pp. 111-31.

US Bureau of Labor Statistics [2000]. Safety and Health Statistics, Survey of Occupational Injury and Illnesses: Public Query Data (1989-current); Selective Access (pre-1989). [http: //www.bls.gov/oshhome.htm]. Date accessed: October 30.

US Department of Agriculture [1918]. Fire prevention and fire fighting on the farm. Farmers' Bulletin No. 904. Washington, DC: Government Printing Office.

US Department of Agriculture [1992]. 1992 Census of agriculture. Washington, DC: Government Printing Office.

US Department of Agriculture [2001]. Census of agriculture. Washington, DC: Government Printing Office, pp. 10-11.

US Department of Health and Human Services [2000]. Healthy people 2010: understanding and improving health. 2nd Ed. Washington, DC: US Government Printing Office.

US Department of Labor [1974]. Occupational injuries and illnesses by industry, 1972. Bulletin 1830. Washington, DC: Government Printing Office.

US Office of Management and Budget [1987]. Standard industrial classification manual. Washington, DC: Government printing Office.

US Office of Management and Budget [2000]. Principles for determining costs applicable to grants, contracts, and other agreements with educational institutions. OMC Circular A-21. [www.whitehouse.gov/omb/circulars/a021/a021.html]. Date accessed: April 15.

US Secretary of Transportation [1971]. Agricultural tractor safety on public roads and farms. Washington, DC: US Government Printing Office, Superintendent of Documents.

Uttal WR [2000]. The war between mentalism and behaviorism. Mahwah, NJ: Lawrence Erlbaum Associates.

Veazie B [1999]. Foundation principles: keys to success of a behavior-based safety initiative. Prof Saf, *44*(8): 24-6.

Waller JA [1985]. Injury control: A guide to the causes and prevention of trauma. Lexington, MA: Lexington Books.

Wallerstein N [1992]. Powerlessness, empowerment, and health: implications for health promotion programs. Am J Health Promotion, *6*(3): 197-205.

Wallerstein N, Bernstein E [1988]. Empowerment education: Freire's ideas adapted to health education. Health Educ Quart, *15*(4): 379-94.

Wallerstein N, Weinger M [1992]. Health and safety education for worker empowerment. Am J Ind Med, *22*: 619-35.

Wallston KA, Wallston BS, DeVellis R [1978]. Development of multi-dimensional health locus of control (MHLC) scales. Health Educ Monographs, *6*: 160-70.

Wandersman A [2001]. Community mobilization for prevention and health promotion can work. In: Schneiderman N, Speers MA, Silva JM, Tomes H, Gentry JH, eds. Integrating behavioral and social sciences with public health. Washington, DC: American Psychological Association, pp. 231-47.

Wardle NJ [1958]. Fatal accidents of Iowa farm people: 1947-1956. AE-870. Ames, IA: Iowa State College Cooperative Extension Service.

Weinger M, Lyons M [1992]. Problem-solving in the fields: an action oriented approach to farm worker education about pesticides. Am J Ind Med, *22*: 677-90.

Weinstein ND. [1993]. Testing four competing theories of health-protective behavior. Health Psychol, *12*(4): 324-33.

Weinstein ND, Rothman AJ, Sutton SR [1998]. Stage theories of health behavior: conceptual and methodological issues. Health Psychol, *17*(3): 290-9.

Weiss CH [1997]. How can theory-based evaluation make greater headway? Eval Rev, *21*(4): 501-24.

White CJ [1916]. Certain occupations as contributing factors to disease of the skin. Boston Medical Surg J, *175*(2): 35-45.

Whitman SD, Field WE [1995]. Assessing senior farmers' perceptions of tractor and machinery-related hazards. J Agr Saf Health, *1*(3): 199-214.

Wholey JS, Hatry HP, Newcomer KE [1994]. Handbook of practical program evaluation. San Francisco: Jossey-Bass Publishers.

Wickes GW, Williams WA, Guthrie FE [1979]. Exposure of field workers to organophosphorous insecticides: sweet corn and peaches. Arch Environ Contam Toxicol, *8*: 175-82.

Wilkerson I [1988]. Farms, deadliest workplace, taking lives of children. The New York Times, September 26, p. A1; p. A21.

Wilkins III JR, Englehardt HL, Rublaitus SM, Crawford JM, Fisher JL, Bean TL [1999]. Prevalence of chronic respiratory symptoms among Ohio cash grain farmers. Am J Ind Med, *35*: 150-63.

Williams DL [1983]. Iowa farm and people characteristics and agricultural accident occurrence [Dissertation]. Ames, IA: Iowa State University, Department of Agricultural Engineering.

Witte K, Givens VK, Peterson TR, Todd JD, Vallabhan S, Becktold MG, Stephenson MT, Hyde MK, Plugge CD, Jarrett R [1993]. Prevention tractor-related injuries and deaths in rural populations: Using a persuasive health message framework in formative evaluation research. Int Quart Community Health Educ, *13*(3): 219-51.

Wood WB [1963]. Safety — A team approach. Extension Service Rev, *34*(3): 43, 60-1.

Yoder AM, Murphy DJ [2000]. Evaluation of the farm and agricultural injury classification code and follow-up questionnaire. J Agr Saf Health, *6*(1): 71-80.

Yoder J, Watson M, Benson WW [1973]. Lymphocyte chromosome analysis of agricultural workers during extensive occupational exposure to pesticides. Mutat Res, *21*: 335-40.

Young CW [1996]. A brief history of the National Institute for Farm Safety: fifty years of professional growth 1945-1995. Columbia, MO: National Institute for Farm Safety, Inc.

Zearing L [1947]. Farm Safety – Historical Data. Summary of historical events between January 1926 and December 1947. Compiled December. Photocopy.

Zhou C, Roseman JM [1994]. Agricultural injuries among a population-based sample of farm operators in Alabama. Am J Ind Med, *25*: 385-402.

Zwerling C, Burmeister L, Reynolds S, Jones M, Lewis M, Popendorf W, Scarth R, Whitten P [1998]. The Iowa farm family health and hazard surveillance project. J Agr Saf Health *Special Issue, 1*: 13-20.

Zwerling C, Daltry LH, Fine LJ, Johnston JJ, Melius J, Silverstein BS [1997]. Design and conduct of occupational injury interventions studies: a review of evaluation strategies. Am J Ind Med, *32*: 164-79.

General Index

Index of Names